GEOGRAPHICAL EDUCATION IN SECONDARY SCHOOLS

Norman J. Graves

Professor of Education
University of London Institute of Education

THE GEOGRAPHICAL ASSOCIATION
343 FULWOOD ROAD
SHEFFIELD S10 3BP

ISBN 900395 63 X

Printed in Great Britain by
John Wright & Sons Ltd., at The Stonebridge Press, Bristol BS4 5NU.

Contents

Illustrations

Figures

Plates

Tables

Acknowledgements

The Geographical Association would like to thank the following organizations and individuals for permission to reproduce material in this handbook.

Aerofilms Ltd (Gate Studios, Station Road, Boreham Wood, Herts, WD6 1EJ) for Plates 10 and 12.

Barnaby's Picture Library (19 Rathbone Street, London W1P 1AF) for Plate 13.

La Documentation Française (29–31 Quai Voltaire, 75340 Paris) for Figs. 4.14, 4.15 and 4.16 taken from *La Documentation Photographique*, No. 6028 (1977): *Le Bassin Parisien*.

Electricité de France (68, rue du Faubourg Saint-Honoré, 75008, Paris) for Fig. 4.19.

Information and Documentation Centre for the Geography of the Netherlands (Geografisch Instituut van de Rijksuniversiteit, Heidelberglaan 2, Utrecht) for Fig. 4.23 taken from the IDG guide *Zuyder Zee/Lake Ijsell*, 1976; Figs 5.2 and 5.3 taken from *Compact Geography of the Netherlands*, 1974; and Plate 19.

Japanese Information Centre (9 Grosvenor Square, London W1) for Plates 1 to 9.

Kodansha Ltd (2–12–21 Otowa, Bunkyo-ku, Tokyo 112) for Fig. 4.3 adapted from E. Fukui (ed.) *The climate of Japan*, 1977.

Longman Group Ltd (Resources Unit, 33–5 Tanner Row, York YO1 1JP) for Fig. 6.1 taken from *Data in Geography: Cities Unit. Land value, land use and affluence: London, Birmingham, Bristol and Liverpool*, 1978.

Mission d'études pour l'amenagement de la basse vallée de la Seine (3 rue Adolphe Cheruel, 76000 Rouen) for Fig. 4.17.

N. J. Graves for Figs 7.1 and 7.2 taken from *Curriculum planning in geography*, 1979; and Plate 14.

N. J. Graves and J. T. White for Figs 4.6, 4.7, 4.8, 4.10 and 4.11, Tables 2 and 3, and the extract on pp. 26–9, taken from *Geography of the British Isles*, 5th ed., 1978.

Netherlands Survey Department (Ministrie van Oorlog, Topografische Dienst, Westvest 9, Delft) for Fig. 4.24 (part of 1 : 50 000 Zwolle sheet).

Oxford University Press for material on the Aswan Dam (pp. 41–3) from M. Simons, *Deserts*, 1967

Photothèque Perrin (28 rue Louis Hubert, 78140 Vélizy, France) for Plate 11.

Royal Netherlands Embassy (38 Hyde Park Gate, London SW7 5DP) for Plates 15 to 18.

Schools Council Geography 16–19 Curriculum Development Project for Fig. 7.3 taken from *Project News*, No. 2, March 1977.

Statistics Bureau, Prime Minister's Office (95 Wakamatsucho, Shinjuku, Tokyo 162) for Figs 4.4 and 4.5 taken from *Statistical Handbook of Japan*, 1978.

Preface

The previous handbook in this series, entitled *Geography in Secondary Education,* was first published in 1971 and subsequently reprinted. When the Publications Committee of the Geographical Association asked me whether I would revise it, I readily agreed, believing that some modifications to the text was all that was required. The more I thought about this, the less happy I became with the prospect, for in the years which had elapsed since 1971 my own thinking had changed, as had ideas about geography and to some extent its role in the education process. I therefore resolved to rewrite the booklet and only to maintain, for reasons of printing costs, two coloured illustrations and the contexts in which they appeared in the previous booklet. Because the text is new I have altered the title to *Geographical Education in Secondary Schools*.

The present handbook begins with a consideration of the curriculum and of geography's function within it. The pluralistic nature of geography as a discipline of knowledge is briefly sketched out and my own preference for an ecosystem view of geography for schools is indicated, as well as its educational aims and objectives. The main part of the book is a consideration of possible teaching strategies illustrated by examples of these. The evaluation of geographical education is then examined and resources for the teacher are indicated. The last section returns to the theme of curriculum by examining a model of curriculum planning for geography.

This handbook is essentially a brief review of what geographical education in secondary schools involves in the nineteen eighties. It cannot go into great details on each issue, but, since the last book was written, a wealth of material has been published to which the interested student or teacher may refer.

I should like to thank the Geographical Association's Publications Committee for its tolerance of my delay in re-writing this booklet; Mr Derek Buttivant and Miss Joan Williams who kindly commented on the draft; Mr Kenneth Wass of University College London, who drew the maps and diagrams; Mrs Teresa Tunnadine for allowing me to incorporate her oil refinery game and Ms Ann Barham whose attention to detail in seeing the manuscript through the press and whose diligence in seeking permission to use photographs and maps I greatly appreciated.

NORMAN J. GRAVES

1 The Curriculum Problem

Any teacher is bound to be faced by a curriculum problem. What is he to teach, how is he going to teach what he has decided to teach and with what resources, and how is he going to evaluate what he has taught. In many societies, what he is supposed to teach is decided for him by the central education authority which sometimes also prescribes the nature of the evaluation to be undertaken. In Britain, teachers in any one department may decide what they wish to teach, granted this is acceptable to the head teacher and the local education authority. In practice, it is assumed that teachers are professionally competent and the local education authority seldom interferes in the details of a school's instructional programme. Nevertheless, it would be wrong to assume that the nature of the school's curriculum is uninfluenced by society outside the school. Societal pressures may be exerted through parental influence; through the school governors, some of whom represent the local electorate; and through the influence of the Department of Education and Science, which itself may reflect the attitudes of the Government or lobbies acting upon the Government. Thus the curriculum in general is likely to respond to various societal pressures and geography as part of the total curriculum is also subject to the forces of change. It will be part of the purpose of this handbook to examine the contribution of geography to the secondary school curriculum and indicate some ways in which the geography curriculum may be put into practice. In so doing it will be useful to examine the kind of pressures affecting both geography and geography in education. First let us examine the concept of curriculum.

1.1 CURRICULUM

In a loose sense, the curriculum of a school can be described as the kind of activities which go on in that school with a view to achieving its educational purpose. That is, the curriculum appears as a means of achieving the overall aims of the school. If we attempt to analyse the idea further, we shall see that this "means" of education consists of a group of teachers interacting with students using a wide variety of subject matter, which is referred to as the content of education. Thus the curriculum may be conceived of as a dynamic interaction process in which the aims and objectives of the teacher, the subject content he uses, the kind of teaching strategies he employs, and the methods of evaluation he devises, all play a part as shown on Figure 1.1.

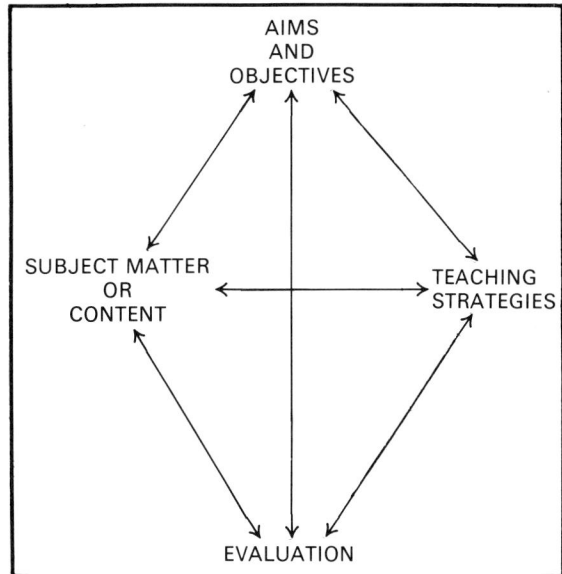

FIGURE 1.1. Simple model of the interactive curriculum process

Clearly the teacher must know what he is attempting to get his students to learn, so he must have certain objectives. Equally, he cannot achieve any learning without a body of content

1

or subject matter which the students may examine and work with. He also needs to be clear as to how he wants the students to work with the subject matter, so he has to devise a teaching strategy or method. Lastly the teacher will want to know how his students have fared, and so he will want to evaluate the learning and the value of the activities suggested. But each element in the situation is capable of influencing every other element. For example, although it would seem logical to begin by having a clear objective in mind, in fact this objective may be suggested by the existence of an aerial photograph in the teacher's resources which shows the layout of a marina and holiday resort in Southern Greece (Plate 14), which enable him to teach the skills of aerial photograph interpretation as well as make elementary points about the influence of physical factors on land use.

Similarly, the subject matter itself may suggest an appropriate teaching strategy, as when industrial location problems suggest the use of a simulation or simulation game. Again the evaluation may lead to a modification of subject matter, teaching strategy and even objectives, if the learning experience has been unsuccessful. The teacher may judge that the objective has been inappropriate, the content too difficult and the teaching strategy uninteresting to the pupils. Hence the dynamic interactive nature of the curriculum process.

It must be admitted that the term curriculum is also used in a more static sense, as when it implies a list of the subjects taught at school. In this sense it is used to indicate the content aspect of the curriculum process. Thus when the curriculum of a school is said to consist of English, mathematics, natural science, history, geography, religious education, physical education, and so on, it is suggested that certain areas of knowledge form the curriculum, and that for each area there exists a syllabus which lists the topics to be covered. It is a less useful notion of curriculum than that of the curriculum process which emphasizes that the curriculum is always in a state of flux, responding to changes in subject matter, in objectives, in teaching methods and in evaluation. A school curriculum which is static will rapidly get out of date and no longer fulfil the needs of the students or of the society in which the school is situated. Thus to go on teaching about the "cotton textile industry in Lancashire"

or about the "cotton belt" in the southern USA as though these were paramount educational objectives is to have become anchored to a content which may have been relevant 40 years ago but which has not great import today.

1.2 FORMULATING THE CURRICULUM STRUCTURE

Since the curriculum can be viewed as a means of achieving the educational aims of a school or college, it follows that it is important for these aims to have been clearly formulated. It is not appropriate in a book on geographical education to devote much space to a long discussion of the aims of education, especially as these are adequately discussed in other publications.[1] It is necessary to state nevertheless that most educators agree that schools have a dual purpose as agents of society. First, they have the task of attempting to develop young minds so that these are awakened to the intellectual feast to be derived from the world of ideas and to the joys of aesthetic experience. In other words, schools have the duty to attempt to unfold each student's potential for learning to its maximum. Secondly, they have a responsibility to society to teach students that which will enable society to continue to function, albeit in a changed form. That is, there needs to be some connection between what is taught in schools and the needs of society in economic and social terms. These needs were spelled out in the 1977 Government Green Paper.[2] Broadly this indicated that pupils and students should be made aware that the country's well-being economically depends not only on the level of technology in use but also on our willingness and ability to work with this technology. This itself depends partly on attitudes and partly on the level of scientific and technological education. Political education was also required to make a liberal democracy work; students should be made aware of how our political institutions work and how they may be changed. The quality of life within the country also depends on people's awareness of the environment, how it is changing and whether these changes are likely to enhance or reduce its quality. This again presumes education for environmental awareness and the setting up of criteria for environmental quality. The environment, however, consists not only of the natural and built environment, but also of the social environment. It is important that the nature of social structure and social relations be understood by pupils and students.

Whilst such an education process for the development of mind and for the social, economic and political needs of society, should be an ongoing one throughout a person's life, schools have the special responsibility of undertaking this for the rising generation. The problem for schools is to decide how best to organize the curriculum to achieve the broad aims postulated above. Traditionally in secondary schools, the curriculum has been of the "collection" type; that is, it has appeared on the timetable as a series of subjects like English, mathematics, French, physics, biology, woodwork, history, geography, home economics, religious education and so on. To some extent this has reflected the way individual disciplines have developed over time in higher education or the perceived needs of individuals in society. Thus geography began to appear on secondary school timetables in the late 19th and early 20th centuries and became widespread as the subject grew and developed in universities. Home economics or domestic science as it used to be called, was placed on the timetable as it was felt that certain pupils should be taught the essentials of running a home. More recently the "collection" type curriculum has been challenged, particularly by sociologists,[3] who feel that such an arrangement of the curriculum creates rigidities in education which hamper its progress and development as well as leading to the creation of high status and low status curriculum subjects for different groups of pupils. The alternative favoured is a much looser arrangement of the curriculum in which the containment of subjects within narrow bounds is discarded for what is generally known as an "integrated" curriculum.

Although a number of experiments have been carried out with integrated curricula in secondary schools,[4] in practice the whole of the secondary school curriculum has not been concerted to date (1979) into an integrated whole. What appears to be the case is that in some 11–18 comprehensive schools, the first two or three years of the curriculum has been organized in such a way that history, geography, social studies and sometimes religious education and English have been combined into a "humanities" course or a course with a name like "integrated studies" or "combined studies" or even "world studies". There is no standard pattern, neither is there necessarily any common ground between the various sources. Many schools still teach individual subjects, and in the upper years of the secondary schools geography remains one of the subjects taught in most schools.

The problem then becomes that of devising ways in which the subjects, and geography in particular, may be used to fulfil the general aims of education. Even if we consider the "integrated" curricula of the lower secondary school, we still need to know what contribution geographical knowledge can make to it. This involves us in considering briefly the nature of modern geography.

2 The Nature of Modern Geography

2.1 SEMANTICS AND EPISTEMOLOGY

A discussion on the nature of modern geography is likely to run into difficulties unless certain semantic and epistemological problems are recognized in the first place. Let us clear up a relatively simple question of word meaning. Although the Greek origin of the word geography literally means "writing about the world", it is quite clear that no word can maintain its original meaning in all its pristine purity. Meanings are modified and elaborated over time. The word geography no longer just means a description of the world, it has evolved its meaning to include an area of study and research concerned with the explanation of certain phenomena on the earth's surface. We shall examine the nature of the phenomena and explanations later. It should also be clear that meanings differ according to the culture in which they are used. For example, it must not be assumed that geography means the same thing to a Briton and to an American. What we call physical geography (geomorphology and climatology) would tend to be called "earth science" in the USA and departments of geography in American universities are seldom concerned with research in pure geomorphology or climatology. Similarly it must not be assumed that what is taught under the heading of geography in French or German schools will correspond exactly with what is taught in British schools. Even among the citizens of one country the nature of the meaning of the word geography will vary: the view of geography as an inventory of physical features, economic resources and political divisions is not yet extinct among the population at large.

Given that modern geography is a body of knowledge yielded by research activity, there remains the problem of classifying this knowledge within the totality of human knowledge. As we saw in the first chapter, the "collection"-type curriculum is based on the assumption that certain school subjects derive directly from disciplines which exist in higher education. These disciplines are differentiated from one another by certain characteristics and provide problems and methods of research which are assumed to be worth pursuing. Hirst,[5] using a reductionist approach, has attempted to divide knowledge into seven fundamental "forms" based on the nature of the concepts used, the nature of the linkages between concepts and the tests for truth used to validate the knowledge. Geography is not included in the fundamental forms as Hirst argues that geography has no fundamental concepts and methodology of its own, but like architecture, medicine and education, it is a "field of knowledge" concerned with problems which use the fundamental forms of knowledge in whatever combination seems suitable. This is not the place to contest Hirst's case in detail,[6] suffice it to say that it is not generally accepted by philosophers concerned with this epistemological problem, and that in any case Hirst argues that though the "fundamental forms" of knowledge should be represented in the curriculum they need not be taught as separate subjects. More acceptable is the view proposed by King and Brownell,[7] that disciplines emanate from the work of communities of scholars or researchers and that though these focus their attention on certain related problems, processes, phenomena or institutions, the work undertaken is not limited by clearly defined criteria. Thus though the community may recognize at any one time what Kuhn[8] has called a research "paradigm", that is to say, generally acceptable objectives and research methods for the discipline, the work may nevertheless overlap with that of another community of scholars. For example, geographers and economists overlap in their work on industrial location, and physicists, geographers and mathematicians overlap in their work on understanding the general circulation of the atmosphere. From this discussion, we should not expect geography to be a subject which stands out as clearly differentiated from neighbouring disciplines, neither should we expect it to have a unique methodology. Its grounds for knowledge are similar to those of the natural and social science; that is, it is vested in both empirical and theoretical research in so far as it is concerned with establishing general principles or theories; that is, in its nomothetic form. In its idiographic form, its grounds for knowledge may be found in phenomenology: in the intuitive reactions of those practising the art of interpreting the essential meanings of places.[9]

2.2 BRIEF HISTORICAL PERSPECTIVE

The recent history of geographical endeavour has been relatively eventful; consequently it may be useful to review this briefly. Any scientific discipline tends to move through a series of stages in which at first it begins to gather information or make observations on the phenomena in which its scholars have an interest. Thus for a very long time the output of geographical work was essentially a collection of facts about the nations and areas of the world. Gradually interest focused on the classification of the data, such as Herbertson's attempt to divide the world into natural regions, Köppen's classification of climates and Vidal de la Blache's division of human activities into *genres de vie* or styles of living. Then began the phase of seeking explanation for phenomena. In practice these phases overlapped. The simplest explanatory theory was that suggested by Ratzel but developed by Semple and Hungtington, namely environmental determinism. Here human activities were looked upon as essentially determined by certain physical factors such as relief, soil and climate. Thus geography became the science which explained human occupations in terms of their determinants in the physical environment. Vidal de la Blache and his followers saw such a theory to be simplistic and developed a style of explanation which took into account the

influence of historical and cultural factors as well as of physical factors on human activity in a given area. This kind of explanation was essentially historical and functional and tended to be associated with the description of the complicated web of interrelationships which existed in a region. Many geographers in the inter-war years espoused this kind of work and saw geography's essential purpose as that of characterizing the personality of a region. This became known as the landscape or regional paradigm of geography. Its effect in schools was, with environmental determinism, predominant up to the 1960s. In the 1950s in the United States and Sweden, geographers expressed dissatisfaction with the regional paradigm as it was seen as a dead end, incapable of generating new ideas or research beyond the detailed description of yet another region. Consequently attempts were made to develop a scientific geography whose purpose would be to express in terms of theoretical relationships, the kinds of spatial regularities which could be observed on the earth's surface, such as the distribution of urban areas or central places. This kind of scientific human geography was made evident in the UK by such books as *Frontiers in geographical teaching*[10] and *Models in Geography*.[11] The change from the regional to the spatial organization paradigm came to be known as the conceptual revolution in geography.

It is important to bear in mind that those geographers who had been undertaking research in geomorphology and climatology were not directly involved in the conceptual revolution. Their paradigm of research had always been based on the scientific method and their objectives had always been to develop theory about the processes of physical landscape change or about those of meteorology. The nature of the human geography which resulted from the conceptual revolution is best illustrated in Abler, Adams and Gould's book *Spatial organisation: a geographer's view of the world*[12] in which the kinds of problems dealt with are those of location, of networks, of land use, of urban fields of influence, and of spatial diffusion. The ultimate aim was to produce theory which could predict and therefore be of use to those taking decisions about locations, networks, land use and so on. This kind of geography was within the philosophical tradition known as logical positivism, and was in harmony with much of the work done in the physical and social sciences. It had a considerable impact on school geography in the 1960s and 1970s in Britain and many A-level, O-level and CSE syllabuses were devised with that kind of geography in mind.

2.3 THE PRESENT POSITION OF GEOGRAPHY

In retrospect, the conceptual revolution seems relatively clear, even if at the time trends were not so obvious and the debates about the nature of geography were heated and protracted. Inevitably those established practitioners in schools and universities who found that their ideas were being subverted reacted in what Schon[13] has called dynamic conservatism in order to attempt to maintain the status quo. Some ignored the conceptual revolution as though it had not happened. Nothing, however, stands still and developments in geography have continued to occur which have placed scientific human geography in a new light. Whereas the spatial organization paradigm of geography is neutral in its commitment to society, the "welfare approach" adopts a more committed approach with social justice as its main aim. Thus where scientific geography is supposedly value-free, that is it examines and theorizes on what spatial relationships exist in society, welfare geography is normative in its approach as it makes recommendations about spatial policy based on the assumption that certain outcomes are desirable, such as distributive justice. In Britain, David Smith's *Human Geography: a welfare approach*[14] is a good example of a book expounding this view. Some geographers are committed to a Marxist form of explanation in human geography. Their views have been gathered together by Richard Peet in *Radical Geography*.[15] Although they do not speak with one voice, their broad argument is that the ideals of social justice can never be realized in a society where the "social relations of production" are those of a capitalist society in which the means of production are privately owned. The Marxian analysis applied to urban geography is illustrated in David Harvey's *Social Justice and the City,*[16] in which he sees "urbanism" as the modern expression of the "social relations of production" in an urban society which is now more powerful

than industrial society, but within which arise the inevitable contradictions of Marxian theory; for example, the confrontations between inner city versus suburb. How an urbanism based on exploitation can be replaced by one appropriate for human beings is a task which radical geographers would like to chart and undertake.

Parallel with the development of scientific geography, there has grown up a literature which has some affinity with the idiographic geography of the inter-war years, as well as with the conceptual revolution of the 1950s and 1960s. This is the literature concerned with what is broadly called environmental perception, which at one end of the spectrum attempts to generalize about the mental images that people have of an environment, as in the studies of farmers' perceptions of drought hazards,[17] and at the other end is concerned with the highly personal place images of individuals and their intimate meanings, as exemplified in some of the works of Relph[9] and Yi-Fu Tuan.[18] Much of this work is of great significance to teachers since it is concerned with explaining how people view the world they live in and the kind of cognitive maps they hold.[19]

2.4 CONCLUSION

Modern geography is not a monolithic subject easily described in a few words. Rather is it a "focused curiosity", in which the world as the home of man provides the focus, but in which the curiosity strays in many directions. The concepts of space and place provide another focus for investigation through scientific and non-scientific methodologies. Since geography is studied in institutions of higher education, these to a large extent determine the kind of knowledge which is created and written. At the present time it is arguable that British geography has the following branches:

1. An earth science branch which is concerned with explaining patterns and processes in biogeography, geomorphology and climatology and therefore using the methods and techniques of the natural sciences including systems theory.
2. A social science branch which is concerned with explaining patterns and processes in human geography, and is dependent on the methods and techniques of the social sciences.

Within this branch some geographers would claim to be committed to certain social objectives.

3. A phenomenological branch which is concerned with eliciting the meaning of certain spatial experiences. The adherents to this branch would claim that their work is complementary to scientific geography and not to be seen as rivalling it. There is a growing literature in this field though at present (1979) it is relatively undeveloped.

Modern regional geography is in effect human geography applied to a particular region, as for example when the problem of regional development in Southern Italy is considered. The use of quantitative techniques, whether mathematical or statistical, is an integral part of much research in both earth science and human geography. These techniques are merely tools which must be chosen with care to suit the problem under investigation. Some geographical relationships may be expressed mathematically, and in such cases, mathematics is merely a symbolic alternative, and more economical language than words. Some geographers have tried to find an overarching framework within which most of what is done in the name of geography could be encompassed. The most appropriate of such frameworks is probably the notion of ecosystem, since this is an organizing concept which subsumes a host of others. It includes the notion of the interdependence of all elements within the system, and existence of small subsidiary systems within the global system. It may be represented diagrammatically as shown on Figure 2.1.

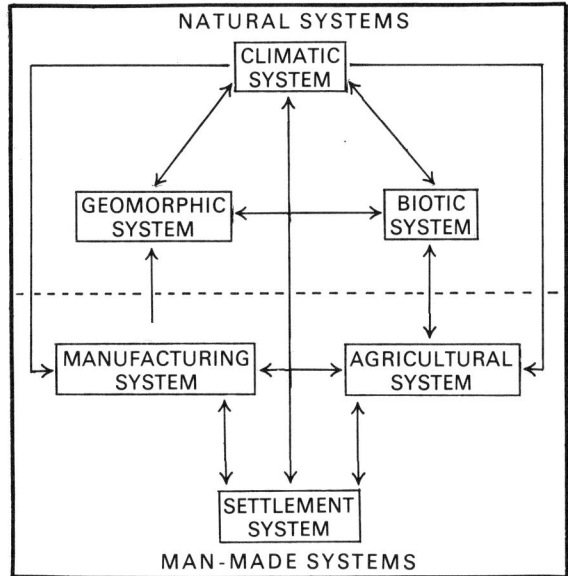

FIGURE 2.1. Simple ecosystem model of geography as a discipline of knowledge.

Like all models, this model of geography is a simplified representation of a very complex reality, and therefore cannot be used to cover all aspects of present-day geography. For example, it is not obvious how the phenomenological aspects of geography can be covered by the ecosystem approach which is essentially scientific. Nevertheless it has the merit of being clear, of showing the relationships between human and physical geography and therefore may be commended for use with upper secondary school students especially those in the sixth form.

3 Aims and Objectives

In Chapter 1 we saw that the curriculum process involved an interaction between aims and objectives, teaching strategies, subject matter and evaluation. We did not then attempt to define precisely what we meant by aims and objectives, except to examine what schooling was meant to achieve and what the Department of Education and Science thought should be the broad aims of education. From this it follows that aims in education are general and necessarily broad in scope. They indicate the general direction in which teachers should be guiding their pupils and students. They are long term and difficult to evaluate for achievement. Objectives on the other hand are more specific and relate to a short-term situation. They are concerned with the immediate targets for a lesson or teaching unit. They can be evaluated so that teachers may find out whether students have learnt what they were meant to learn. Thus to aim at making students aware that there may be cultural bias in their view of foreigners is a long-term aim, whilst to teach a class that travel across 15° of longitude is likely to cause a change in local time by one hour is a short-term objective.

3.1 AIMS IN GEOGRAPHICAL EDUCATION

In so far as a subject like geography is taught either as a separate subject or as part of a combined subject group, there is a need for its practitioners to be aware of the contribution that the subject is making to the education of the pupils. It is this awareness on the part of the teacher which enables him to give direction to his day-to-day teaching. The general aims of the subject give the necessary criteria which enable the teacher to choose his more specific objectives. Without these the teacher could set up a series of teaching units which though they had specific objectives did not necessarily contribute in a rational way to the education of the students. For example it would be possible to have a series of essentially factual or informational objectives which did little to develop the student's mind or general cognitive perspective.

What then are the aims of geographical education? First it must be stated that these, like all aims, are value judgements and it may not be possible to obtain general agreement about these. Secondly, the aims of various subjects are bound to overlap to some extent. Geography teachers, like other teachers, will reinforce the work of their colleagues in English and mathematics and contribute to the development of literacy and numeracy among the students, and more especially render them more able to handle graphic information. As Balchin has argued,[20] geography is a particularly good vehicle for the development of graphicacy among students, given the subject's preoccupation with spatial relations expressed through the medium of maps, block diagrams, graphs, aerial and ground-level photographs, and remotely sensed imagery like the landsat images. Thirdly, geography has a special responsibility for the development of what has come to be known as mental maps or the development of spatial conceptualization. We all live and work in a three-dimensional environment and we need to operate effectively within it. This means that not only do we need to have a mental image of our immediate home environment, but we need accurate mental images of our routes to frequently visited places. Further we need to understand how to get to places that we have never visited, and we often need to understand how things are related to one another in space even though the things are not yet present, for example in designing a house and locating it within a piece of ground already partially occupied by other objects. Clearly this ability to develop mental maps is closely related to graphicacy. Once a mental map has been represented on a document, then the ability to use that document is graphicate ability. The development of mental maps depends on many factors, such as a student's experience, but field-work of all types in which close observation of the environment is undertaken is potent in developing such mental abilities. Fourthly, geography is particularly concerned with problems which have an important spatial component. Thus one of the important aims of geographical education is to develop a tendency in students to examine the spatial aspects of social and economic problems. Industrial location, town planning, migration of population, traffic in towns and between towns, land use, the diffusion of innovation; all these are topics which have spatial elements and about which problems arise requiring solutions. They are topics which are both spatial and relevant to the students' lives.

Fifthly, geography can contribute significantly to environmental education. Environmental education[21] is a broad term which covers all aspects of education whose aims are to develop in students a heightened awareness of the total environment and a caring attitude about the environment. The need for environmental education became apparent in the 1930s when resources of timber and minerals as well as soil were being consumed at an alarming rate. At that time the emphasis was on conservation education. However, it soon became clear in the post-war years that the rapid development of economies in all parts of the globe were having a deleterious effect on the environment as a whole. Waste gases were polluting the atmosphere whilst the rising discharge of effluent in rivers and seas were causing the Rhine and the Mediterranean, for example, to become lifeless. Further, the concept of environmental quality was being extended to the cultural landscape both in rural and urban areas. Consequently environmental education is concerned both with the equilibrium of the ecosystem in the biosphere and with the aesthetic aspects of man's environment. Geography's special contribution to environmental education lies in the development of environmental awareness through direct observation in the field or indirect observation through secondary sources. It may also help, with other aspects like biology, to develop positive attitudes to the environment with a view to a commitment to environmental quality. But this presumes that teachers will be able, with the help of parents and other interested adults, to develop the notion of a balanced and aesthetically pleasing environment. Some prob-lems, like the control of effluent are clear enough. Others like the concept of harmonious building development need development over time. Fortunately, some of the materials from recent curriculum development projects help the teacher in this direction.[22]

It is necessary to emphasize that these general aims are for the teacher. It is doubtful whether the pupils and students would fully understand these as justifying the teaching of geography or any other subject. Most students are highly pragmatic in their attitude to school knowledge; they tend to look upon such knowledge as a means to an examination or to a job, even if incidentally they may become interested in what they are studying. Lastly, the very long-term nature of these aims makes them difficult to verify. They must therefore constitute an article of faith among teachers of geography rather than a proven justification.

3.2 OBJECTIVES IN CURRICULUM PLANNING

A first step in devising a teaching unit for implementation in the classroom, is the formulation of a fairly precise and specific objective. As we saw in Chapter 1 this objective may be suggested by a resource the teacher has, by an examination question set in the previous year, by an aspect of geography like "air drainage" which appears relevant, or indeed by a teaching strategy like group discussion which the teacher wants to try out with an appropriate topic such as "Should a by-pass be built round the village of Capel on the A24?" Such specific objectives will have meaning to both students and teachers since they will aim at realistic and achievable targets such as "The student shall be able to identify a depression on an isobaric weather chart and its associated warm and cold fronts and indicate the probable extent of the rain belt associated with the warm front". Assuming this to be aimed at an O-level 5th-form group (16-year-olds), it is concerned with both the skill of reading a weather map and the understanding of the relationship between a warm front and a rain belt. It is both a skill objective and a cognitive objective. The problem arises as to whether such teaching objectives should be stated in the form of behavioural objectives. The idea of behavioural objectives

stems from the work of such psychologists as Skinner who affirm that one can only
tell what people have learned from the kinds of
behaviours that they exhibit. For example, if
someone can draw a cross-section across a valley
from a 1 : 10 000 map of a country district where
contour lines are wide apart, this is evidence that
he can draw that cross-section, but not evidence
that he can draw any cross-section. Consequently
the purists would require that objectives should
specify: the precise behaviour indicating that the
objective has been achieved; the product by
which the achievement can be evaluated, e.g. a
cross-section; the conditions under which the
behaviour is performed, e.g. with what map,
what ruler, what paper, during what time; the
standard to be used to evaluate the achievement,
e.g. what errors are allowed in the cross-section
to qualify for achieving the performance. As may
be seen, specifying objectives of this nature is
elaborate and time consuming. Teachers are
probably too much under pressure to consider
the detailed specifications of behavioural objectives in their day-to-day teaching. There are also
solid objections to such objectives. They assume
that it is possible to so control the learning of
individuals, that the result of teaching can be pre-
specified in great detail. This is empirically
impossible. There are many unintended as well
as intended outcomes to learning situations. In
any case, as Sockett[23] points out, no specification
of behavioural objective is really so completely
unambiguous that teacher and pupil give the
same meaning to the words used.

There is a need, therefore, to be able to state
objectives in a way that makes them reasonably
clear and yet not involve the teacher in an
elaborate ritual of stating those in behavioural
terms. Marsden[24] has suggested that these be
stated in the form of principles. A principle can
be thought of as a general statement linking two
or more concepts and which if understood by the
student enables him to use it in new situations.
Thus the idea that the higher the land the greater
the precipitation it receives is such a principle
since it links up the concepts of altitude and
degree of precipitation. A principle in human
geography would be the notion that the larger a
town the greater the number of its functions and
the larger is its field of influence. The advantages
of using such principles is that they can be
assessed against the criteria specified in section

3.1 on general aims to see whether they measure
up to these aims and they give an indication of
what should be contained in a teaching unit. It
may help to contain a teaching unit within a
reasonable length and prevent it from over-
extending itself into areas not directly relevant to
the principle enunciated. The essential structure
of a school's geography syllabus could be made
up of such principles, though each teaching unit
would necessarily contain all the examples and
resources necessary to enable the students to
learn the principle.

So far objectives have been considered as
statements of intent; that is, of what the student
should learn, and the examples given have all
been in the area of cognitive learning. Bloom[25] in
his taxonomy of educational objectives attempted to devise a hierarchy of such cognitive
objectives from knowledge, comprehension, application, analysis, synthesis to evaluation, in
which knowledge was the lowest in the hierarchy
and evaluation the highest. This taxonomy is a
handy way to evaluate a cognitive objective to
determine at an early stage whether the objective
proposed is a high or low-level one. Clearly all
pure information comes into the lowest category,
whilst objectives which involve students in
assessing the validity of a theory come into the
highest category. In general it may be stated also
that up to the age of 14 or 15 most students will be
able to tackle objectives involving knowledge,
comprehension and application, but they will
find difficult objectives which involve analysis,
synthesis and evaluation. The taxonomy is not,
however, a precise instrument and it cannot be
used to specify in detail the level of difficulty of a
particular objective. For example, though comprehension is low in the hierarchy, as most
teachers will readily testify, much depends on
what the students are being asked to com-
prehend. A statement about the nature of the
different districts in a town may be easy to
understand; one outlining the process which led
to the differentiation of these districts would be
much more difficult to understand.

Objectives may also be affective[26] as well as
cognitive; that is they relate to emotions and
attitudes rather than to knowledge. Thus in the
area of environmental education, the teacher is
often more involved in changing attitudes than in
communicating knowledge. In practice it is impossible to teach only to affective or cognitive

objectives. Any learning experience is governed by the hope that the learning will be enjoyable (an affective objective) and that a skill and/or principle will be learned (a cognitive objective).

Lastly, as has been hinted at, objectives are an indication of intent but the teacher can never be sure at the end of a teaching unit that the objectives he has set the class will in fact have been achieved. This may be for many reasons: the objective may have been too ambitious, the learning resources inappropriate, the teaching strategy misconceived, or more simply the class may not have been in a mood to work on that particular occasion. This is a remediable situation, since the teacher, having got feedback from the class, can modify his approach next time. There may, however, be many occasions when the teacher may not want to specify objectives in the form of a principle or attitude to be developed. For example the teacher may wish to stimulate the class to throw out a series of ideas on a problem, without imposing any limitations on the kind of solutions which are acceptable. Let us take a particular instance: suppose money becomes available to a local community to improve the amenities in the neighbourhood. The students may be given information, maps, diagrams, but the action to be taken is wide open. In such a sense, no specific objective is appropriate. Eisner[27] states that such an educational encounter has "expressive" rather than instructional objectives. In effect the objective is to get students to react in a creative way to a problem.

3.3 SUMMARY

In this chapter we have considered the nature of aims and objectives in geographical education. We have seen that the teacher in preparing curriculum or teaching units for his student, must have a clear idea of what he is trying to achieve. Each teaching unit must therefore have certain objectives. We have also seen that these may be stated in great detail or in vague terms, but that an acceptable and useful compromise is to state them in the form of a principle to be learned. It has to be borne in mind however, that all objectives have an affective as well as a cognitive element within them. Without appeal to the emotions, without any motivation, learning seldom takes place. Further, certain teaching units may be designed not to teach a certain principle, but to enable pupils and students to respond creatively to a new situation. Such teaching units are said to have "expressive objectives". Finally it must be remembered that the number of short-term objectives possible in geography is immense. The selection of objectives must be done in the light of the general aims of geographical education. These are value statements about what geography teaching is attempting to achieve in the long run. Besides reinforcing literacy, numeracy and graphicacy, geographical education aims at developing spatial conceptualization and an ability to solve the spatial element of social and economic problems. It also contributes to environmental education.

4 Teaching Strategies

Teaching strategies are an integral part of the curriculum process, but that part about which teachers are often most concerned. In discussing this issue at one time we should have been talking about teaching methods. It is still broadly true that what concerns us is how to put over a particular idea, skill or attitude, but whereas the term "method" was often interpreted in a narrow technical sense, such as the "discovery method", with the suggestion that this or that method was a desirable one to use, the term teaching strategy is more open and suggests that there exists a range of such strategies for teachers to use, depending on the circumstances. Any teaching situation is subject to a great number of variables: the age of the students, the homogeneity or heterogeneity of the class, the background knowledge, degree of motivation and experience and personality of the teacher, the physical nature of the classroom, the resources available to the teacher, the general atmosphere of the school, and so on. Consequently, it would be foolish to state that any one "method" is bound to yield desirable results. Much depends on what one is trying to achieve, with whom and in what circumstances. There is a need to be pragmatic and judge a strategy on its results.

Are there then no criteria for selecting a teaching strategy? We may be able to indicate certain characteristics of teaching strategies which need to be observed. The first is that there ought to be some harmony between ends and means. This may be illustrated by some simple examples. If a teacher is attempting as part of his aims to get students to solve spatial problems such as an industrial location problem, then he must give them plenty of opportunities to undertake problem-solving exercises. To give them ready-made answers to problems will in no way achieve the aims he has in mind. Similarly, the teaching of a skill such as the drawing of

statistical diagrams needs to be reinforced by plenty of practice. The second criterion is that teaching strategies used with any one group should be varied as frequently as possible to offset the possible boredom which may set in if similar techniques are used day in day out. One suspects that the use of worksheets may have had such an effect in the last few years. Thirdly, it is desirable to build into a learning exercise a degree of difficulty which will stretch a student without discouraging him. Whilst we are all conscious of the student who finds work difficult we are perhaps inclined to underestimate the potential of many who can cope.

4.1 A POSSIBLE TYPOLOGY OF TEACHING STRATEGIES

There are many possible ways of classifying teaching strategies, but probably the simplest primary classification is one which depends on whether learning is going on in the classroom or in the field. Within the classroom, one may further subdivide teaching techniques into:

1. Those using varied resources which programme learning activities through a series of steps with a view to the student learning certain concepts or principles. For example, a teacher may lead a class through a series of questions on maps and photographs of our motorway network to obtain the general idea that the motorways form a network linking the main industrial areas of Britain, with nodes occurring in London, West Yorkshire, South Lancashire, Birmingham and Bristol, and that the links pass near but not through other important centres like Leicester and Nottingham. This kind of approach which involves a mixture of oral interaction between the teacher and the class and some written work, often a statement of

the principle to be learned, is a strategy which is in widespread use and was at one time probably the dominant one.

2. Those in which the essence of the objective is still the learning of principles or concepts, but in which the strategy used is that of the case study. For example, the purpose may be to learn that farming can be analysed as a system with inputs and outputs and a feedback element. This may best be done by considering a particular farm as shown in Figure 4.1, where the systems diagram is based on a farm in north-east England.

3. Those in which the essential purpose is the teaching of a psychomotor skill, like cross-section drawing, or a technique of analysis like the rank correlation coefficient. Although the strategy may start by an indication of the purpose of the skill or technique, which may be approached by posing a question, for example, "How can we find out whether there is any general relationship between Gross National Product and crude birth rate in Latin America?", the main thrust of the learning process is focused on the technique which the students are to master. Here the process is usually one of the teacher demonstrating the technique and the students practising it with resources provided by the teacher in the form of maps or numerical data as the case may be.

4. Those in which a problem-solving approach is stressed. In practice many teaching strategies start with the statement of a problem, but in some cases the problem-solving behaviour is that which is concentrated on. The problem to be tackled may be one drawn from the reality of the present-day world, in which the options available are known and limited in number. For example, the student may be asked how he would relieve traffic congestion in his local town's high street. In this case data and maps relating to the problem may be provided for the student to work upon. On the other hand, the problem may be one where the solution

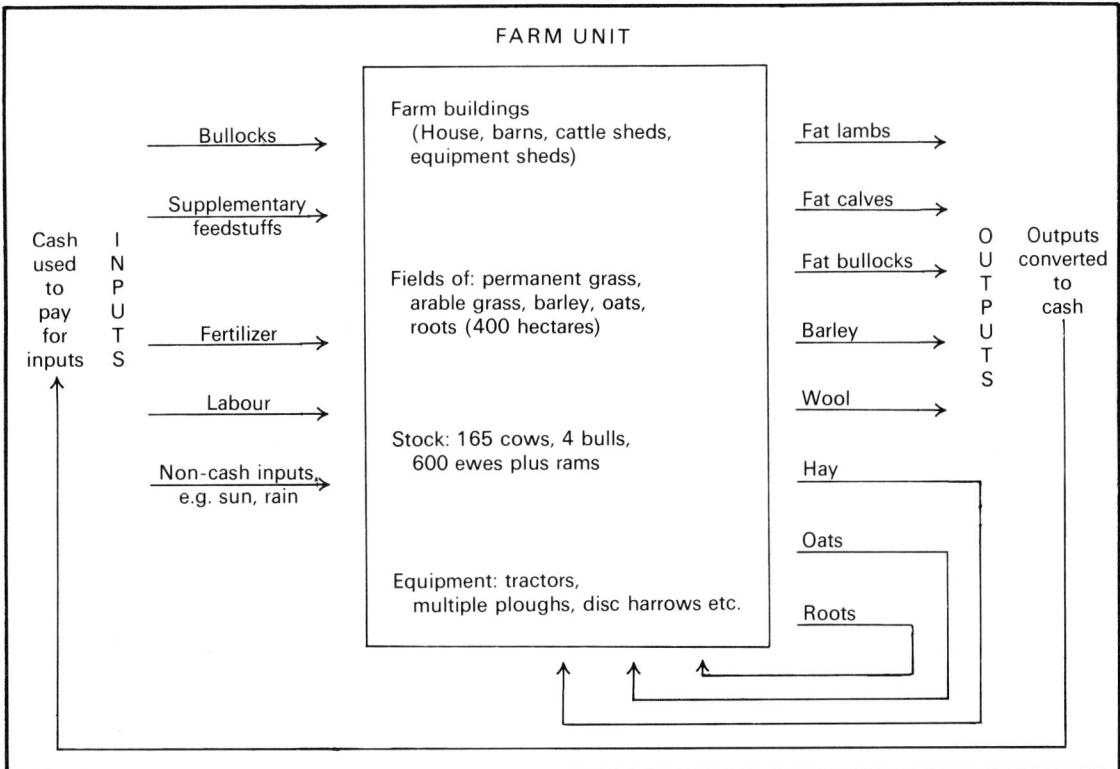

FIGURE 4.1. A farm system based on a beef cattle farm in north-east England

depends on a complicated system of negotiations or personal interactions in which he cannot be directly involved, as, for example, in the problem of deciding on the route of a new motorway, as in the case of the M25 around London. What is important here is not just that the students should be able to work out what might be the cheapest route, or the shortest route, or the least damaging to the environment; but that they should understand how the decision-making process works given the nature of the problem. Here the use of a simulation or role play or a game may be an appropriate teaching strategy, since clearly the students cannot be closely involved in the actual decision making.

In fieldwork one can divide teaching strategies into two groups. The first is the field teaching group in which the purpose of being out in the field (apart from enjoyment) is for the teacher to demonstrate landscape features or physical or economic processes with the "reality" in front of the students. The teacher may also develop the students' power of observation and recording. The second group of teaching strategies are those where the object is to find out something specific about the field area. Again, this may be a survey type of exercise such as a land use survey, or a survey of house types, or a survey of occupations in a village. But such information gathering is by itself not particularly instructive unless there is some sort of purpose behind it. Thus more typically, data is gathered to test a hypothesis, for example that there is significantly more arable land on well-drained light soils than on heavy clay, thereby confirming the important influence of a physical factor on land use; or again one may suspect that the majority of the working population of a village do not in fact work in the vicinity but are long-distance commuters, so gathering data about occupations and place of work may help to confirm or reject the hypothesis. At a more sophisticated level an attempt may be made to find out what factors influenced the majority of the inhabitants to become long-distance commuters.

Fieldwork is an important element of geography since it is, in the ultimate analysis, the means of obtaining basic data from which information may be derived, and from which one must seek evidence for the testing of a hypothesis or for the solution to a problem. In practical terms, fieldwork has become even more important since examination boards have either encouraged its use or even made it a compulsory element of A-level, O-level and CSE courses. This means that teachers need to programme such fieldwork into their year's work. Whilst it is always difficult to make suggestions that will fit every school's situation, it seems possible to argue along the following lines. First, fieldwork should appear in every year of the school course from the first to the upper sixth. Secondly, the amount of time devoted to fieldwork should progressively increase from the first to the seventh year. For a minimum it is suggested that the following pattern be adopted:

1st year.....1 day or 2 half days
2nd year.....2 days (or 1 day + 2 half days)
3rd year.....3 days (or a week-end)
4th year.....4 days (or 1 day + 1 week-end)
5th year.....5 days (or 1 week or other combination)
6th & 7th year. 10–14 days (usually in one or two blocks)

Thirdly, as far as possible, the fieldwork undertaken should be integrated in the geography curriculum, so that it is seen to contribute to the total understanding of geography and complementing what is happening in class. A simple illustration at first form level would be that map-reading skills could be tested in the locality of the school, whilst at sixth form level a series of units on urban morphology and processes may be tested out either in the local town or in a town whilst the sixth form is on a week's fieldwork.

Fourthly, the nature of the fieldwork might progressively change from being simple observational fieldwork in which younger pupils are trained to observe what is considered to be geographically significant, to a much more problem-oriented type of fieldwork in which students are asked to decide what data they need, how they are to treat it and what confidence they may have in the results of that enquiry. To give examples, first formers might be asked to identify simple groups of associated plants on a heathland and observe that this association is much the same over heathlands in many areas and how these are associated with sandy or gravelly soils; this possibly leading to a simple concept of an ecosystem. At the sixth form level, to give a similar example, the students might be asked to investigate the way in which the setting up of a recreation area with a boating and fishing lake, swings and roundabouts, and refreshment kiosks on a heathland may influence the workings of the ecosystem, and to suggest the kind of measures which might be taken to prevent much harm being done to the vegetation as a whole.

Assessing fieldwork undertaken by students for examination purposes involves making separate estimates for (a) the planning of the exercise, (b) the work done in the field, (c) the analysis and interpretation of the results, and (d) the presentation of the work in a report or field notebook. Again circumstances will vary and examination boards may give special instructions on this—as the Joint Matriculation Board does—but I would suggest

30% for planning
20% for work done in the field
30% for analysis and interpretation
20% for presentation.

4.2 PROGRAMMED TEACHING UNITS AND THE USE OF CASE STUDIES

Here the main task for the teacher is to reach a clear-cut objective by getting the students to study and work upon teaching materials which act as the evidence. Let us take a simple case which is illustrated below, namely that of making young children understand why tourism is likely to be one of the main sources of income for people living along Europe's Mediterranean coast. In a simple class and teacher interaction working with an atlas, an aerial slide (Plate 14) and a sketch (Fig. 5.1) derived from the slide, the children can be led to come to the conclusion that tourism is likely to be an important source of income as a result of environmental conditions, as shown below. The layout is in the form of notes for the teacher.

1st form (11-year-old average ability pupils) *40-minute teaching unit*

Objective
    To find out why tourism is one of the main sources of income in many Mediterranean areas.

Resources
    Atlas and Wall Map (if available)

Slide of Greek Peninsula (Plate 14, opp. p. 72)
Sketch of Greek Peninsula (Fig. 5.1, opp. p. 72)

Procedure
1. Refer to holidays and find out if anyone has been to the Mediterranean—if so pupils tell class where it was and what it was like.
2. Tell class we are going to find out why for many places around the Mediterranean tourism is one of the main sources of income.   5 mins
3. Show slide of aerial view of Greek Peninsula (Plate 14).
    (a) What is being shown?
    (b) From what point of view?
4. Analyse picture into its parts.
    (a) Port and pleasure craft.
    (b) Hotel?
    (c) Calm blue sea.   5 mins
5. Conclude on evidence of holiday industry.
6. *Note* by children (either from chalkboard or from duplicated sheet).   10 mins
7. Now show slide again and discuss reasons for holiday industry.
    (a) Would they like to be there? Why?
    (b) Sea.
    (c) Sun (back picture evidence by reference to feel of weather in Mediterranean summer).
        (Could ask children in what season they think this photo was taken—in fact, a winter picture.)
    (d) Absence of farming—no water. 5 mins
8. Children to complete sketch based on photograph.   5 mins
9. Sentences to be written to explain why the holiday industry is an important source of income.   10 mins

This appears inevitably as an example out of context, but it can be part of a series of teaching units on the way people earn a living in a lower school humanities course, or it may be part of a sequence on how people spend their leisure. That particular unit might be extended to consider the impact that tourism has had on the Costa Brava or any other Mediterranean coast, so that man's influence on the environment is illustrated as well as the influence of the environment on man.

The above relates to a situation in which the amount of content used and the essential ideas learned are limited. What follows relates to a higher intellectual level, namely a 3rd form (14-year-olds) situation, considering the problem of population growth in the regional setting of Japan. Again, the format is that of notes for the teacher, including the questions he would ask the class.

**Theme: How Japan has solved her population problem[28]
3rd Year (14 years old)**

Objective

To discover the nature of Japan's population problem.

Resources

35 mm slide projector
Duplicated map of Japan (Fig. 4.2)
Atlases

Procedure

1. Introduce Japan via British trade links (Japanese cars) or other means.
2. Check on position of Japan by following Japan Airlines Flight—London–Tokyo.
3. Check latitudinal and longitudinal reference lines (Atlas and Fig. 4.2).
   (a) What is the nearest line of longitude to Tokyo? (140°E)
   (b) Which line of latitude goes through N Honshu? (40°N)

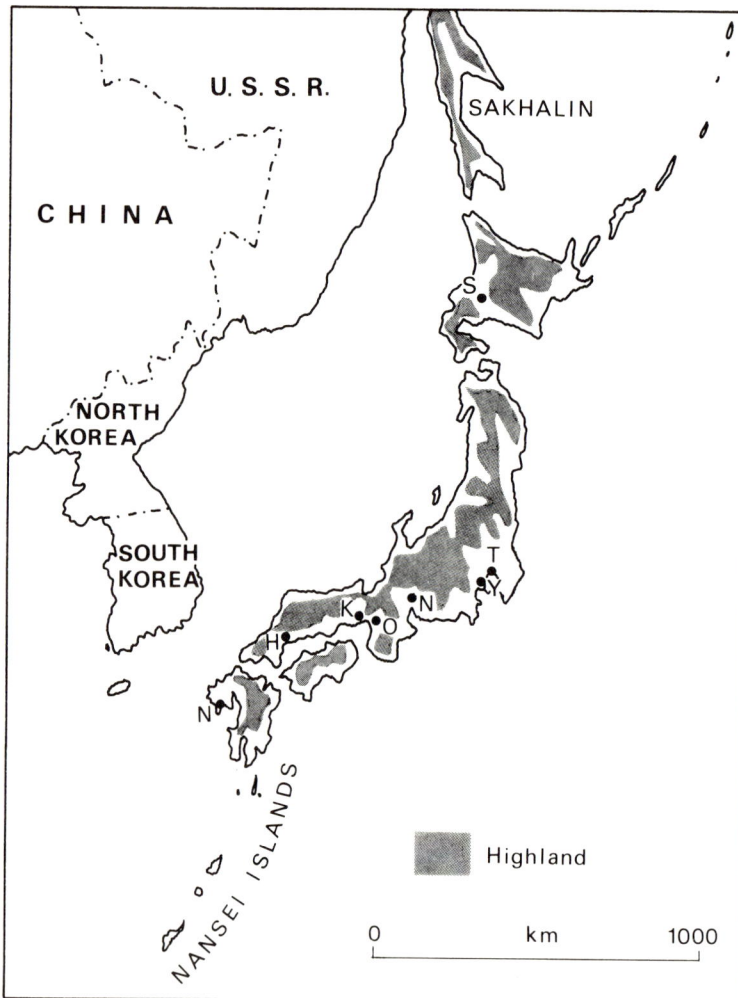

FIGURE 4.2. Map of Japan

(c) How does the position of Japan compare with that of Great Britain? (Further south and a long way east.)

4. Turn to p. – in Atlas (Map of Japan). Duplicated map (Fig. 4.2) to be available.

    (a) In what way is Japan similar to the British Isles? (Group of Islands.)

    (b) From N to S what are these islands? (Hokkaido, Honshu, Shikoku, Kyushu.)

    (c) What are the names of the sea and ocean which surround Japan? (Pacific Ocean and Japan Sea.)

    (d) Pupils to print these on the maps as well as Tokyo and the other towns.

5. In the 1940s and 1950s it was usual for the quality press to have articles on Japan's population problem—we shall try to work out what was meant by this. Look at the following population figures on the screen (Overhead Projector Transparency) (Table 1).

TABLE 1. Population of Japan

Early 19th Century	30 m.
1900	44 m.
1920	55 m.
1930	64 m.
1940	74 m.
1950	83 m.
1960	93 m.
1964	96 m.
1980	117 m.

(a) What was the increase in the 19th century? (14 m.)

(b) What has been the increase from 1900 to 1980? (73 m.)

(c) What has happened? (An acceleration in growth.)

(d) What sort of difficulties arise when a country's population increases rapidly? (Housing, food supplies etc.)

PLATE 1 Winter scene in north-west Japan

PLATE 2   Winter scene in Tokyo

6. However, Japan is not the only country which has had an increase in population. Almost all countries have experienced this. Let us compare Japan and the UK (Overhead Projector Transparency).

	Area km^2	Population millions
Japan	372 000	117 m
UK	245 000	56 m

What do you notice? (Japan about $1\frac{1}{2}$ times the size of the UK, but population over twice that of UK's.) Therefore, the size of the total population in relation to the area may be one problem. But area may not be the only thing to consider; e.g. Netherlands is half the size of Scotland and has twice the population and the standard of living is higher in the Netherlands.

7. Could the climate be a limitation on food production? Let us examine some evidence:

PLATE 3   Summer scene near Tokyo

Project the following slides on the screen:

(a) Winter scene in N.W. Japan—snow clad terraces (Plate 1). Ask for description of scene—what conclusions do you come to about winter weather? (Cold in N.W. Japan with some humidity.)

(b) Winter scene: Tokyo—school band playing (Plate 2). What does weather appear to be like? (Dry and sunny.)

(c) Summer scene: Near Tokyo (Plate 3)—what indications are there of the weather? (Hot—ripe cereal.)

(d) Rainfall map: is there a large area with less than 1000 mm of rainfall? (No.) Where does most of the rain fall? (S. and W. Japan.) (Fig 4.3) Bearing in mind previous photographs—when is the dry season in S. Japan? (Winter.) Therefore climate suitable for agriculture.

Distinction may be made between:
Cold winters in the N.
Warm winters in the S.
Rainfall mainly in summer, but more than adequate.

8. Now if we examine land use figures we find that only 16% of the land is classed as arable. What does this mean? (Small amount of land which can be ploughed, limited food production.)

Let us examine some scenes in Japan to find out the reason.

Project following slides

(a) Kobe (Plate 4). What do you notice about the layout of the city? (Restricted between mountains and sea.) This is Kobe—locate this town on your atlas.

(b) Mt Fuji (Plate 5)—What is this? (A volcanic cone.) What does this suggest about Japan's land surface? (Uneven

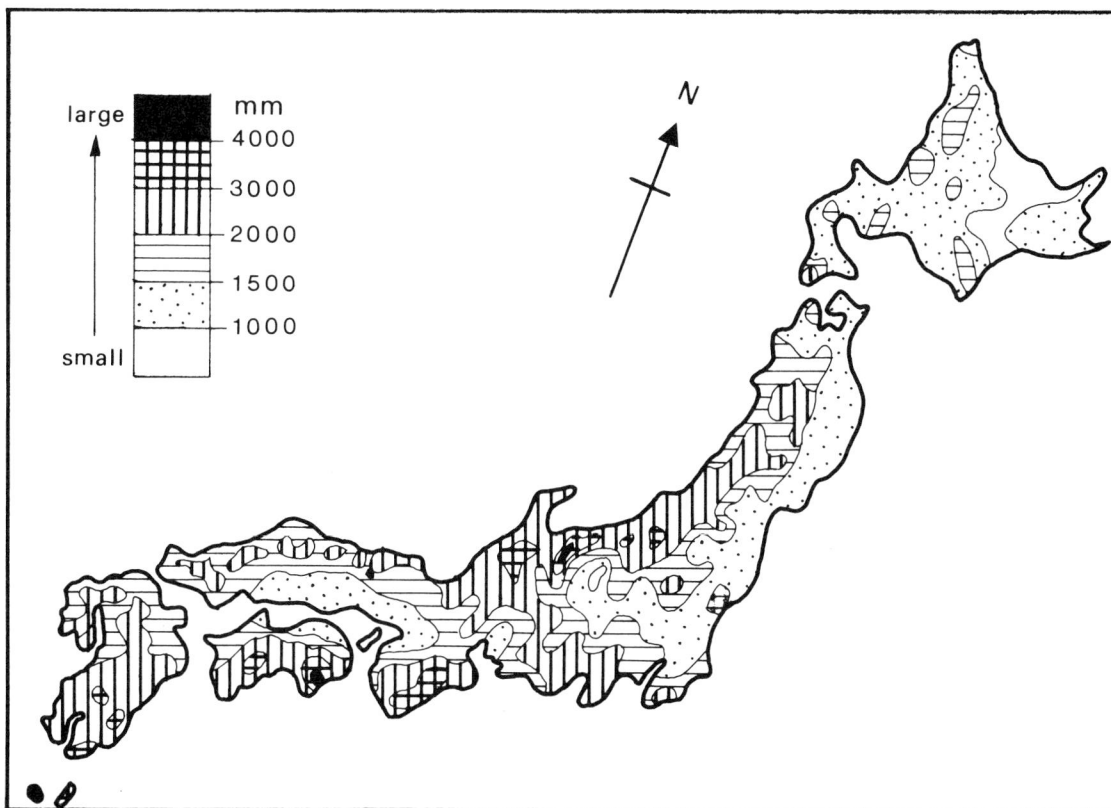

FIGURE 4.3. Mean annual precipitation in Japan

PLATE 4   The town of Kobe

PLATE 5 Mount Fuji in springtime

relief and volcanic activity.)
9. Study your atlas map of Japan.
  (a) Where is there any lowland? (Around coast.)
  (b) Where in particular are the largest areas? (Near Tokyo, around Nagoya and in N. Kyushu.)
  (c) Give these areas the names of Kanto plain, Nobi plain and Tsukushi plain.
10. Print these names on your outline maps and complete the following statement:

Japan

Japan was said to have a population problem because:
(1) (Population increase)
(2) (Need to produce food)
(3) (Limited area compared with UK)
(4) (Relief)

Theme: How Japan has solved her population problem

Objective

To find out the ways in which Japan has tackled her population problem.

Resources

35 mm slide projector
Duplicated questionnaires
Atlases

Procedure

1. Recapitulation on previous unit to establish
  (a) size of population;
  (b) area of country;
  (c) percentage of arable land.
2. We no longer read articles about Japan's population problem, for unlike India's, it is

PLATE 6   Reclaimed land in Japan

no longer such a problem. Let us attempt to find out why.

3. One method of extending the area of arable land on steep slopes was shown in the previous lesson; incidentally—what was it? (Terraces.) (Plate 1) Project slide of reclaimed land (Plate 6). Here is another method of extending the land area.

    (a) What do you notice about the relief of the land? (Very flat.)

    (b) What regular features do you notice? (Straight edges, rectangular plots.)

    (c) What can you see in the middle ground? (Port area.)

    (d) How do you think this area of flat land came about? (Reclaimed from sea.)

    (f) Is this likely to be a cheap and rapid way of increasing the land area? (No.)

    (g) Compare Dutch experience in the Delta Plan.

4. Project slide of motor cultivator (Plate 7).

    (a) What is the farmer doing? (Appears to

PLATE 7   Cultivating paddy fields
          with a motor cultivator

be churning up mud with a mechanical
cultivator.)

(b) What sort of crops is he likely to sow?
    (Rice.)

(c) Now this by itself will not increase the
    amount of land for food production but
    what might it do for each area of land?
    (Might increase production of rice.)

In fact, good techniques plus the intensive
application of fertilisers have raised the
yield of rice per hectare and these are now
three times the yields in India.

(d) Can you suggest why the Japanese
    employ these motor cultivators rather
    than tractors and farm machinery?
    (Size of fields.) An average farm is
    scarcely 1 hectare.

(e) Does the fact that more and more
    machinery is being used, suggest
    anything   about   agricultural
    manpower? (Not too plentiful.)

5. Now there is something else the Japanese
   might do to their agriculture to increase
   food production—look at the following
   figures   (Overhead   Projector   Transparency):

Crop	Area 000 hectares	Production 000 tons
Rice	3 301	12 419
Other grain Crops	1 520	3 925

(a) What does this table show? (Area under
    crops, output.)

(b) What crop seems to give the greatest
    yield per hectare? (Rice.)

(c) Therefore, what would be one way of
    increasing the production of food?
    (Grow more rice.)

(d) Is there a lot of scope for this? (Limited,
    especially when it is realized that with
    the rising standard of living, more
    vegetables and fruit are being grown.)

6. Class to answer part I of Questionnaire.

Questionnaire: Part I

(1) What steps have the Japanese taken to
    increase the area of arable land?

(2) Why are these ways of increasing arable
    land insufficient to meet Japan's food
    shortage?

(3) What have the Japanese done to in-
    crease the productivity of the existing
    arable land?

(4) Why is mechanization of agriculture in
    Japan different from that in England or
    America?

(5) (a) What crop is the most productive
        in Japan?

    (b) Would it be possible to increase
        food production by devoting
        much more land to this crop?

PLATE 8 Industrial
activity in Japan:
petrochemical works

PLATE 9   Industrial activity: Shipbuilding

7. Clearly the extent to which Japanese can
   grow more food for the growing population is limited. Since they cannot grow
   enough, what must they do? (Import.)
8. Project table of imports (Fig. 4.4).
   (a) Does this confirm that food is im-
       ported? (Yes.)
   (b) But what are really the main imports?
       (Crude oil, raw materials.)
   (c) What does this suggest about Japan's
       economy? (Industrial.)
9. Project pictures of industrial activity.
   (a) Petrochemical works (Plate 8). What
       do these look like? (Petrochemical
       works.)
   (b) Shipbuilding activities (Plate 9)
       What is happening? (Shipbuilding.)
       This is at Nagasaki; locate this town.
   What general impression do these two
   slides give of Japanese industry? (Heavy
   industry.)
   (c) Export graph (Fig. 4.5)
       What other goods does Japan produce
       besides machinery? (Radio receivers,
       cars, optical instruments etc.)

10. Therefore, in what way has Japan really solved her food production problem? (By buying food abroad and selling industrial goods in exchange.)
11. This policy has been so successful that the Japanese standard of living has increased enormously in the past 10 years. Associated with this increase in the standard of living has been the following change (Overhead Projector Transparency):

Year	Birth Rate In Japan
1947	34·3/1000
1979	15/1000

Therefore, is there likely to be continued population pressure in the near future?
12. Complete Part II of the Questionnaire.

Questionnaire: Part II
For many years the Japanese have had to .... food to supplement home supplies. The most important other imports by value, however, are ...., ..., wool and iron ore, These are the raw materials of such industries as petroleum refining (Yokohama), textiles (Osaka), ... (Nagasaki) and .... (Kobe). Lighter industries have also developed such as the making of .... As a result of all this industrial development, the standard of living in Japan has increased and this has been accompanied by a .... in the birth rate. Thus population pressure in Japan is considerably .... than it used to be.

Again, such a teaching unit (which would probably take between three and four 40-minute teaching periods depending on the class) could be given in the context of a theme on population, in which different aspects of the population demographic transition could be illustrated.

At the 5th form (16-year-old) level a teaching unit on the changing industrial composition of a given region could be along the following lines. It is based on material contained in a currently available textbook.

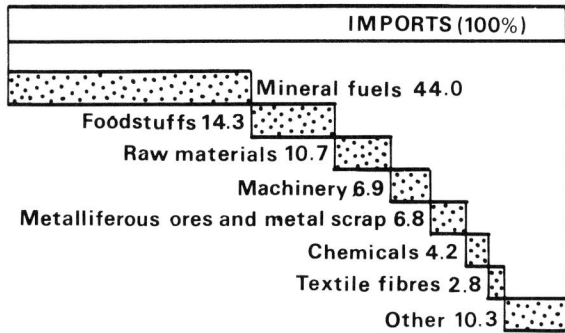

FIGURE 4.4. Japanese imports by commodity group

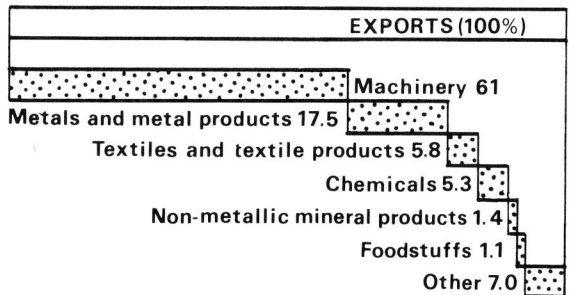

FIGURE 4.5. Japanese exports by commodity group

**West Midlands: Unit I
5th Form (16 years old)**
Objective
To describe and explain the growth of industry in the West Midlands area.

Resources
Photocopied maps
Data sheets
Photocopied extract[29]

Procedure
1. Locate Midlands and check on meaning of traditional region (Textbook or Atlas).
2. Examine map of "Economic Planning Regions" and note division of Midlands into East and West Midlands—our concern is with the West Midlands (Fig. 4.6).

TABLE 2. Economic planning regions of Great Britain

	Area/Economic Planning Region	Population 1976 (thousands)	Area (km²)
	Great Britain	55 928	229 852
I	South East	16 894	27 414
	Greater london	7 028	1 596
II	East Anglia	1 803	12 565
III	South West	4 254	23 660
IV	West Midlands	5 165	13 013
	Metropolitan County	2 743	696
V	East Midlands	3 733	12 179
VI	Yorkshire & Humberside	4 892	14 196
VII	North West	6 554	7 992
	Greater Manchester	2 684	983
	Merseyside	1 578	389
VIII	Northern	3 122	19 349
IX	Scotland	5 205	78 767
X	Wales	2 767	20 763
XI	Northern Ireland	1 538	14 121

Source: Graves, N. J. and White, J. T. *Geography of the British Isles*, 5th ed. London: Heinemann, 1978, p. 37.

3. Look at data on per cent of area of Great Britain and per cent of total population in *West Midlands* (Table 2). Conclude on nature of population density in West Midlands, and possible reasons for this in terms of industrial nature of area. Locate Birmingham and Black Country.
4. Problem: how did the industry develop and change?
   (a) students read extract from *Geography of the British Isles* (see below) and note nature of industrial development;
   (b) discuss implications of modern motor industry for general pattern of industry (see diagram) (Fig. 4.9).

5. Conclusion on the interdependence of industry in the area and its "capital intensive" nature.

The Growth and Development of the Conurbation
The distances by road between Birmingham and several other important towns may be obtained from the chart below (in kilometres). This chart shows one of the advantages of the position of the *conurbation* with respect to the rest of England. This central position however, cannot be the only reason why the Birmingham area developed. Otherwise one might have expected a town like Lichfield to have increased in size to a greater extent than

```
LONDON
315 LIVERPOOL
186 244 BRISTOL
269 205 325 HULL
123 346 120 352 SOUTHAMPTON
630 341 586 392 667 GLASGOW
597 338 584 355 661  70 EDINBURGH
304 117 310  88 358 336 306 LEEDS
246 262  70 358 189 592 590 334 CARDIFF
438 246 456 187 502 229 170 147 478 NEWCASTLE-upon-TYNE
256 117 262  98 307 384 357  53 286 198 SHEFFIELD
176 144 141 197 205 462 461 174 163 322 123 BIRMINGHAM
```

FIGURE 4.6. Economic planning regions and regional capital populations (1976 estimates) in Great Britain

it has in fact done. It would be dishonest to give one single explanation for the growth of Birmingham and the Black Country since all sorts of influences were at work in this area. We can only note how the area gradually grew in importance. Certainly at the time of William the Conqueror, there was nothing more than a rather small manor near the river Rea (a tributary of the Tame) which could be forded at that place. It is possible that owing to the existence of this crossing place many people from the surrounding area came this way and it became profitable to establish *markets* here.

In the early days people required simple articles for farming and for themselves. We know that by 1650 the making of nails, locks, bits for horses, leather harnesses, were well established industries in the area. Later on in

the 18th century, buttons, buckles and trinkets of iron and brass were being made in Birmingham as well as guns, metal being cast in moulds of local sand. Clearly some form of heat energy was required for many of these industrial activities and fortunately there was locally available a *coal seam* which outcropped widely in the Black Country area called the "Thick Coal" because it was often 10 metres in thickness and therefore easy to extract. There were also ironstones, although at first the iron which they made was often not up to the quality required by the manufacturers of metal goods who had to import *pig iron* from elsewhere. However, with improvement in *smelting* processes and the rise in the demand for iron, blast furnaces were set up in the Black Country. Until 1858, 182 had been built and

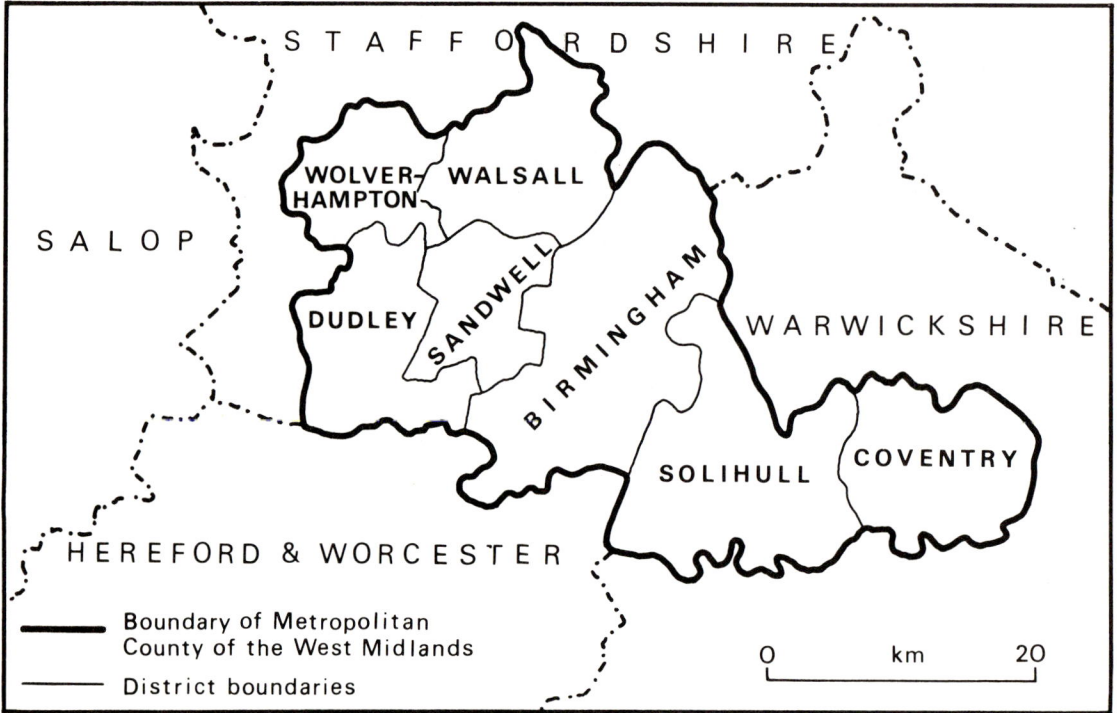

FIGURE 4.7. Administrative divisions of the West Midlands

the Black Country was described as "a great workshop both above ground and below; at night it is lurid with flames of the iron-furnaces: by day it appears one vast loosely-knit town of humble houses amid cinder heaps and fields stripped of vegetables by smoke and fumes" (Mackinder H. J., *Britain and the British Seas*).

The basis of all this industrial development was undoubtedly the *coalfield* which was for long known as the South Staffordshire Coalfield and which now forms part of the West Midlands division of the National Coal Board's areas. The areas where *coal measures* (the rocks containing coal) appear on the surface are marked on Figure 4.8. In fact coal is

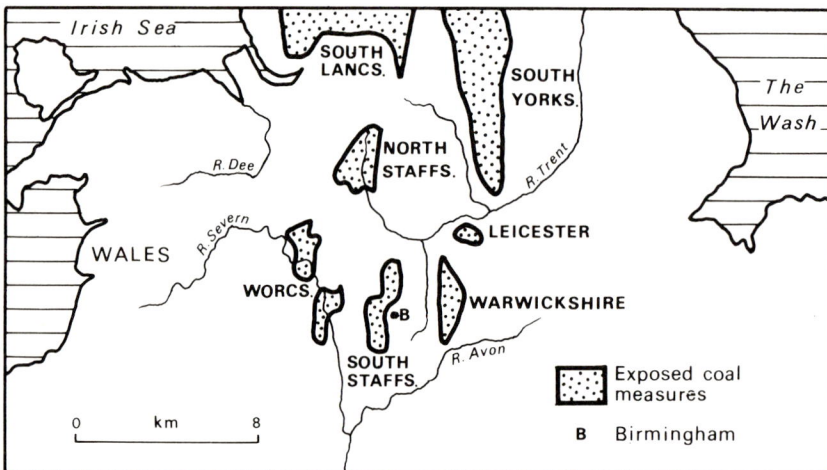

FIGURE 4.8. Coalfields of central England

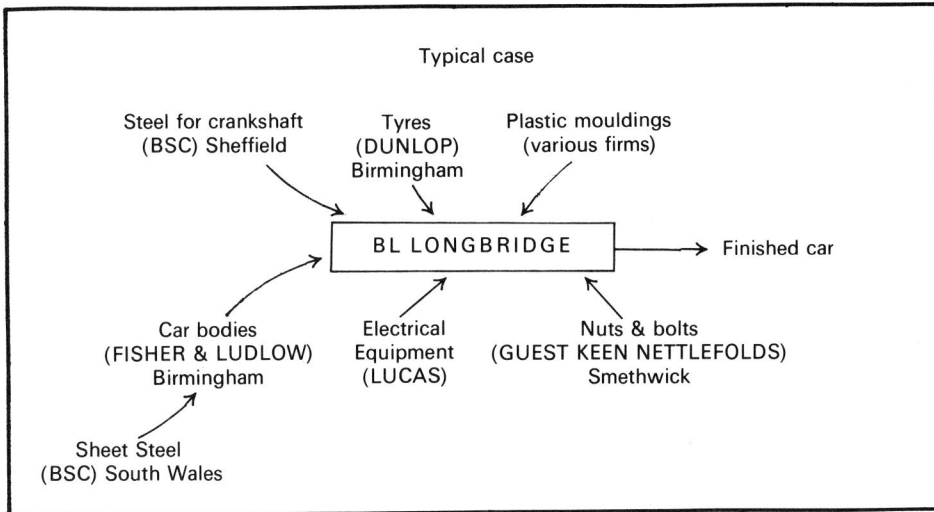

FIGURE 4.9. Motor car assembly in the Midlands

often mined where the *coal measures* are much deeper underground, and today little coal is mined in the exposed coal measures area either because it is not worth exploiting or because it has already been mined.

To help transport the raw materials of industry canals were built in the 18th and early 19th centuries, so that today the area possesses a relatively dense network of old, narrow and little-used canals. Turnpike roads (i.e. paying roads maintained by a private company) were also built and helped the growth of trade and industry in this area. Later, after 1840, railways were constructed. As a result the older workshops and factories were built along roads or crowded together along the banks of canals, whilst late-19th-century industrial growth tended to be along the railways.

With the decline in iron and coal resources, the heavy iron and steel industry tended to contract, thus today the only major unit left in the industry is the British Iron and Steel (Tube Investment) works at Bilston [and that is due to be closed]. But the many metal-using trades remained since the metal used could always be obtained from elsewhere. And as we saw earlier the number of such trades increased to include the cycle, motorcycle, and motor car industry. In all these the value of the metal used is only a small proportion of the cost of the article produced. Most of the modern industries are

capital intensive industries, that is industries in which the capital cost of the equipment used is a major element in the cost of the product, as for example in an automated motor car assembly plant.

(Extract from Graves, N. J. and White, J. T. *Geography of the British Isles.* London: Heinemann, 1978, pp. 56–8.)

West Midlands: Unit II
5th Form (16 years old)

Objective

To demonstrate the need for and some problems of urban renewal, using Birmingham as an example.

Resources

   1 slide (Plate 10)
   2 photocopied maps (Figs 4.10 and 4.11)
   1 photocopied table (Table 3)

Procedure

1. Recall division of West Midlands conurbation into Birmingham and Black Country.
2. Show aerial photo of West Bromwich (Plate 10) and ask for description. Conclude area was a mess landscape-wise.
3. Why should this be so? Discuss unplanned growth and decline of industry.

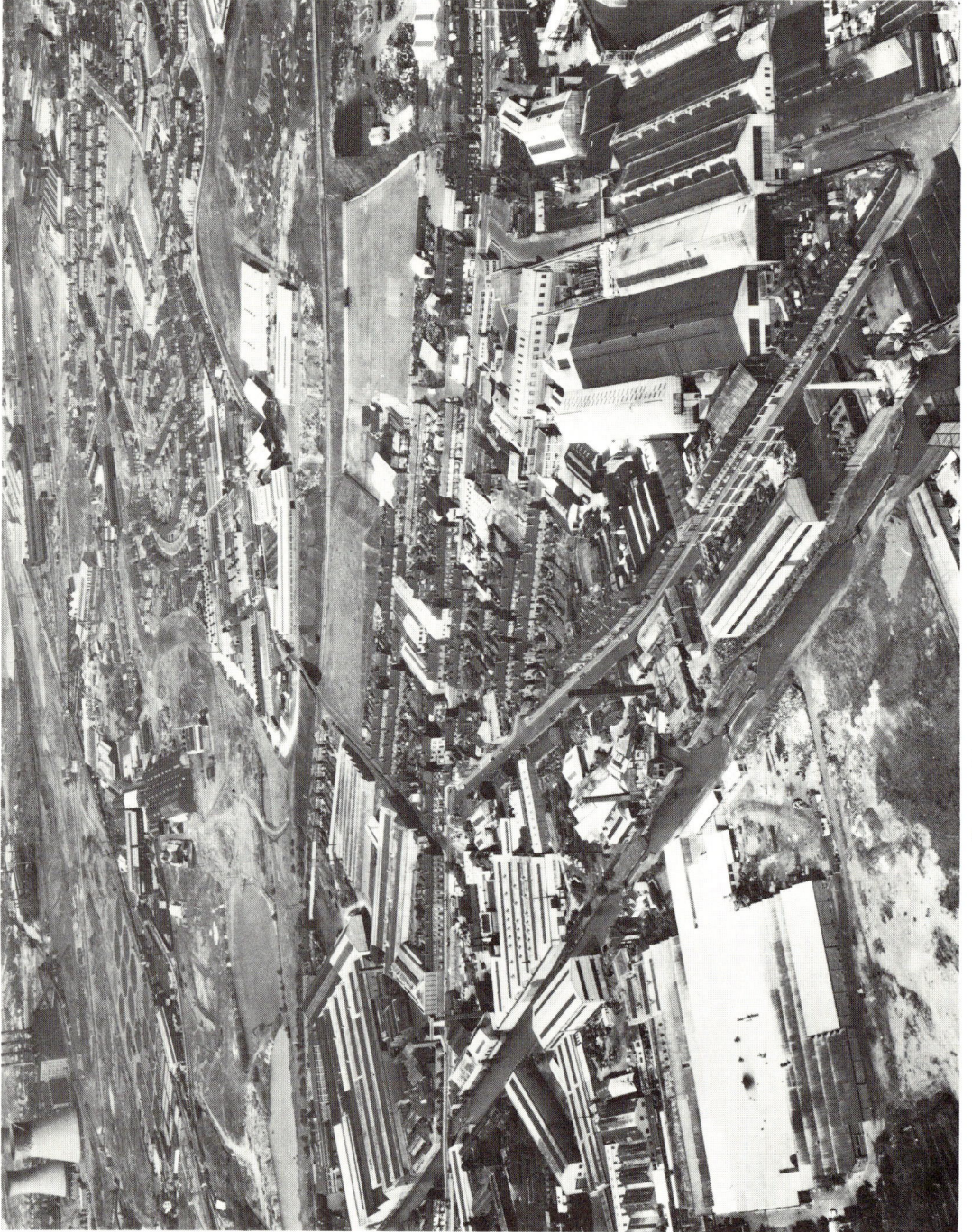

PLATE 10   Derelict land in West Bromwich

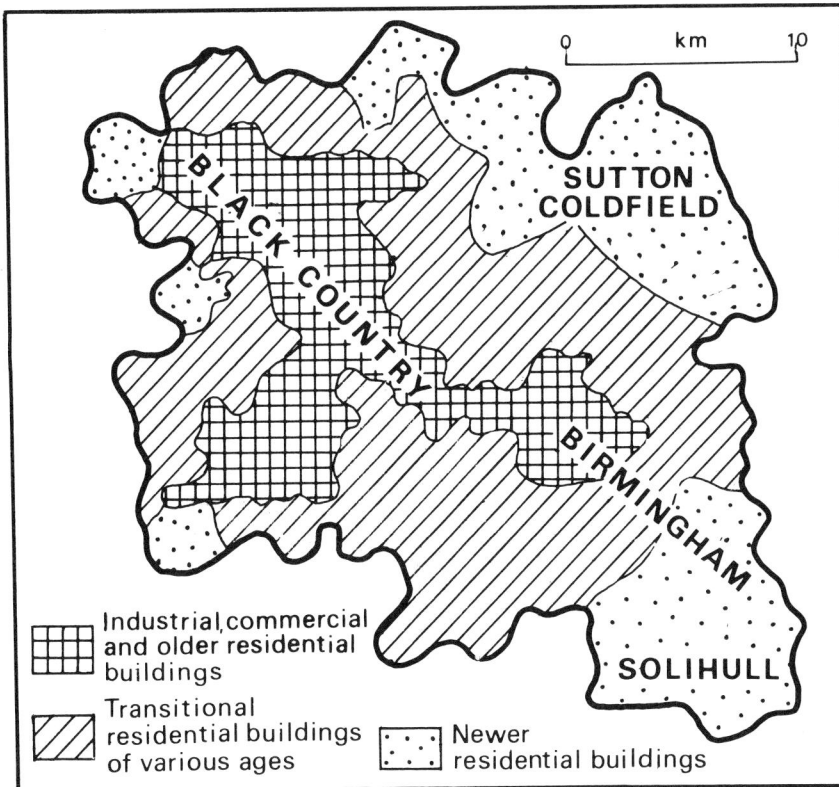

FIGURE 4.10. Land use in the West Midlands conurbation

4. Examine map of land use in West Midlands conurbation (Fig. 4.10).
   What pattern of land use and growth emerges?
   Refer to who lives in the various zones.
5. State post-war planning legislation made redevelopment possible—known as "Urban Renewal".
   Take case of Birmingham.
6. Students to put down their own ideas as to what planners might want to do. Compare with what they actually set out to do.
   Put this on chalkboard.
   (a) Redevelop part of the city for new housing.
   (b) Find sites for schools.
   (c) Increase areas of open space.
   (d) Allocate land for industry.
   (e) Improve the road system.
7. Leaving out the C.B.D., how are they tackling urban renewal? Look at evidence on map of redevelopment areas (Fig. 4.11) (neighbourhood units, land zoning, linking ring road).
8. Look at land use table for development areas—amount of open space + or − ? (Table 3). What does increasing amount of open space imply about type of housing?
9. Examine table of land use in development areas to check on total population + or − ? Therefore, what happens to the others?
10. For homework—work out some of the problems of carrying on such a redevelopment programme.

Another way of using a programmed teaching unit is for individual work by pupils or students, with the teacher acting as a guide and helper, but not orally directing the lesson except in starting the class off. For example, assuming that a teacher wishes a group of 4th year (15-year-old) students to find out the main physical and

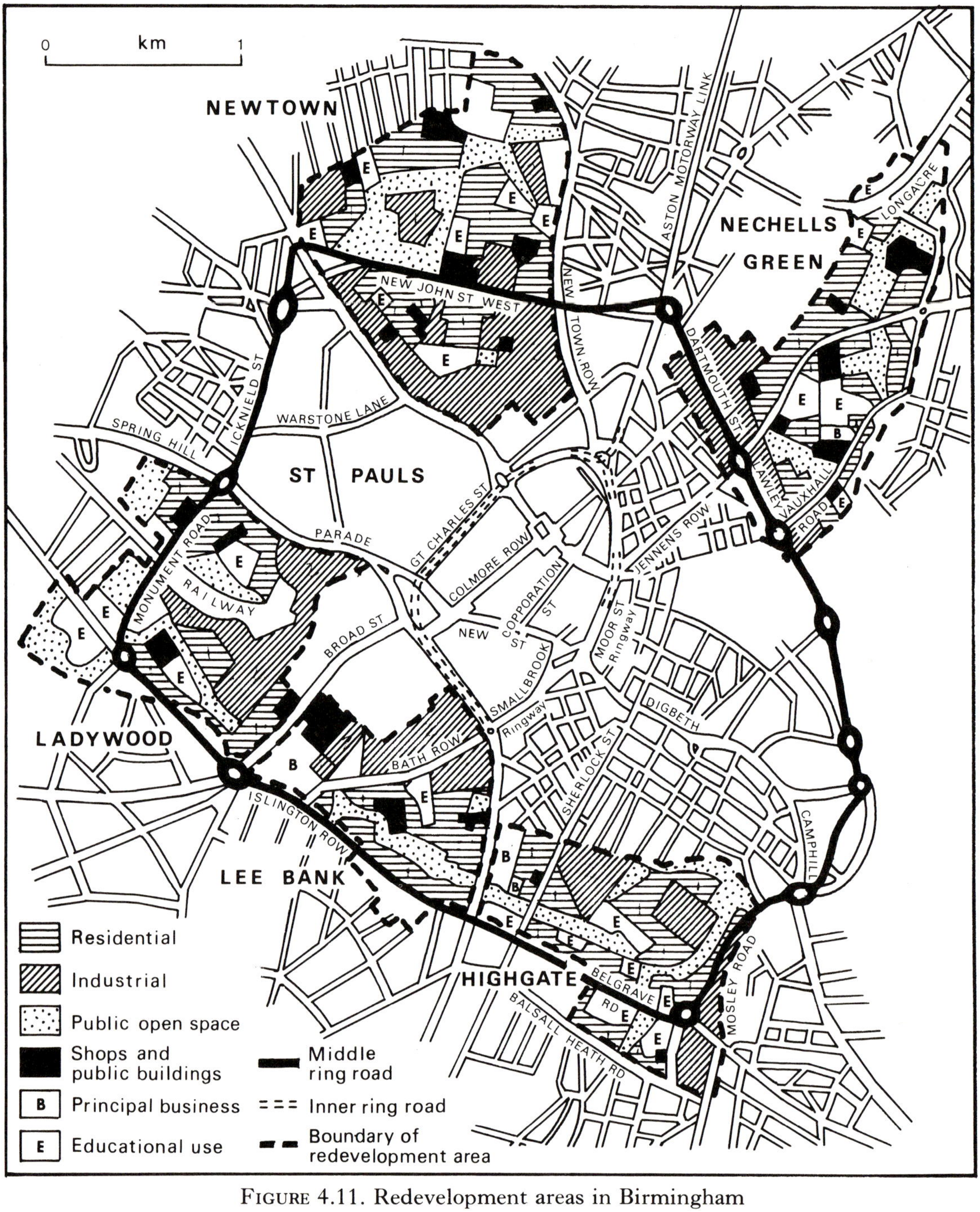

FIGURE 4.11. Redevelopment areas in Birmingham

TABLE 3. Land use in Birmingham's five Comprehensive Development Areas

Area	Total Redevelopment Area (hectares)	Population A (B)	No. of Dwellings A (B)	Education		Industry (hectares) A (B)	Public Open Space (hectares) A (B)
				No. of Schools A (B)	Areas (hectares) A (B)		
Nechells Green	106·8	12 537 (19 072)	3 635 (5 885)	10 (12)	10·8 (3·2)	24·8 (25·2)	16·8 (1·6)
Newtown	159·6	15 400 (28 125)	4 467 (9 349)	14 (15)	16·0 (5·2)	47·6 (47·6)	24·1 (2·6)
Ladywood	115·6	12 448 (24 418)	3 609 (7 558)	9 (9)	10·0 (1·8)	24·8 (24·6)	20·0 (0·8)
Lee Bank	76.8	6 531 (14 797)	1 894 (4 492)	5 (7)	5·1 (1·4)	16·0 (17·7)	10·0 (0·08)
Highgate	94·4	10 080 (16 484)	2 924 (4 886)	15 (8)	12·8 (2·4)	21·4 (21·4)	15·6 (3·9)
Total	553·2	56 996 (102 896)	16 529 (32 170)	53 (51)	54·8 (14·0)	133·2 (136·4)	86·4 (8·8)

Unfit dwellings before redevelopment 24 670

A After Redevelopment
(B) Before Redevelopment

Source: Graves, N. J. and White, J. T. *Geography of the British Isles*, 5th ed. London: Heinemann, 1978, p. 90.

economic factors favouring port and port industry expansion, he might use the port of Le Havre in France as a *case study* and proceed along the following lines. The work sheet is reproduced below.

Port and port industry expansion
4th Form (15 years old)

We shall attempt to find out what kind of physical and economic factors favour the growth of ports and their associated industries. The following instructions will help you to do this. We shall study the port of Le Havre in some detail.

Questions
1. Study the map in Figure 4.12 and state briefly the location of the town and port of Le Havre in France. Approximately, how far is it from Paris in kilometres? How long

would the journey take, by rail or road given an average speed of 80 kilometres per hour? How does that compare with a journey from London to Southampton? (*c.* 110 km.)

Read
2. The growth of the population of the town is given in Table 4 as well as comparisons of its trade in 1952 and 1975.

Questions
3. What has happened to the population and trade of the port since 1952? What product accounts for the greater part of the increase in the trade of the port? What part of the trade dominates, imports or exports?

Read
4. What has happened since 1952 is part of a long history of growth. The port of Le

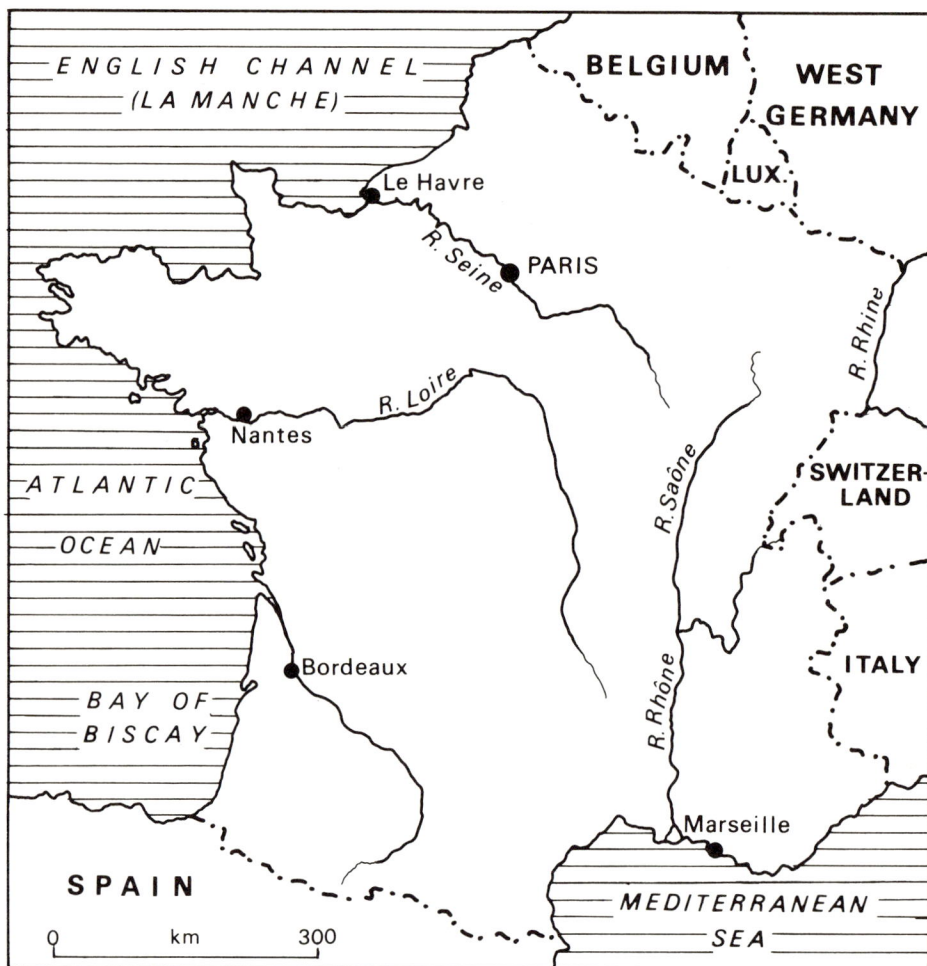

FIGURE 4.12. Map of France to show location of Le Havre

TABLE 4(a). Population of Le Havre, 1936–79

Year	000s
1936	186
1952	157
1962	181
1975	264
1979	269

TABLE 4(b). Trade of Le Havre in 1952 and 1975

Trade	1952	1975
	million tonnes	
Imports	10 062	57 573
(Petroleum)	(8 236)	(50 735)
Exports	3 260	16 307
(Petroleum)	(2 132)	(9 960)

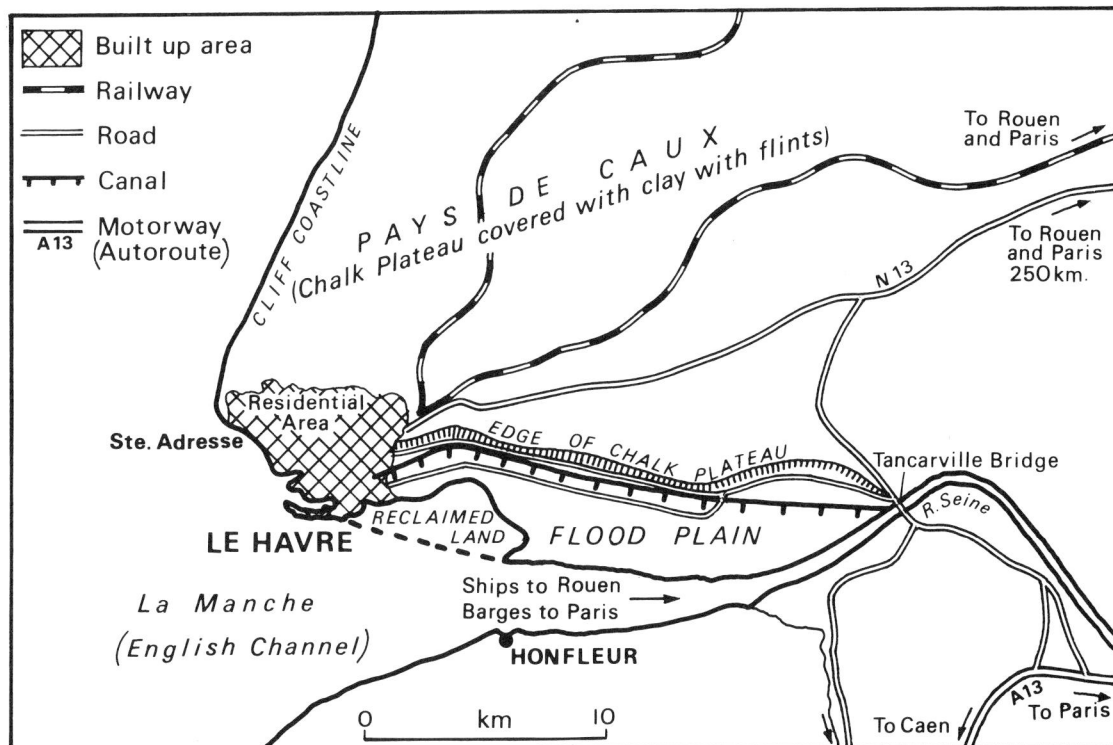

FIGURE 4.13. Position and site of Le Havre

Havre was a creation of King François the first in 1517. The port was needed to replace the small port of Honfleur (Fig. 4.13)[30] on the opposite bank of the estuary of the River Seine. It consisted at first of two small irregular docks. By the 18th century the Bassin du Commerce constituted the main part of the port (see Fig. 4.14). In the 19th century other docks were added such as the Bassin Bellot. By the middle of the 20th century the port is as shown in Figure 4.14 with the Bassin de la Marée as the latest addition. In 1975 the port was as shown in Figure 4.15 and covered about 100 km². It had even grown a deep water outport for giant oil tankers at the Cap d'Antifer as shown in Figure 4.16. Such a growth in the physical size of the port is partly a response to need and partly an anticipation of further development.

Questions

5. Examine the maps (Figs 4.14 and 4.15) carefully and indicate what you notice about the changing size of the dock basins over time. How can you explain this? What made the construction of dock basins relatively easy from a physical point of view? (Fig. 4.13) Only a small part of the recently developed port area is, in fact, used for dock basins—what are the other areas used for? (see Fig. 4.15).

Read

6. The development of industry in Le Havre has gone hand in hand with the growth in the port traffic. Originally the port traffic consisted of such raw materials as coal, cotton, coffee, sugar cane, rubber and wheat; that is, raw materials, many imported from tropical countries, some of which were at the time colonies of France. Exports, always lower in volume and in value, consisted of some engineering products (e.g. cars) and manufactured foodstuffs. In the 19th and in the first 50 years of the 20th century Le Havre was France's main transatlantic passenger port.

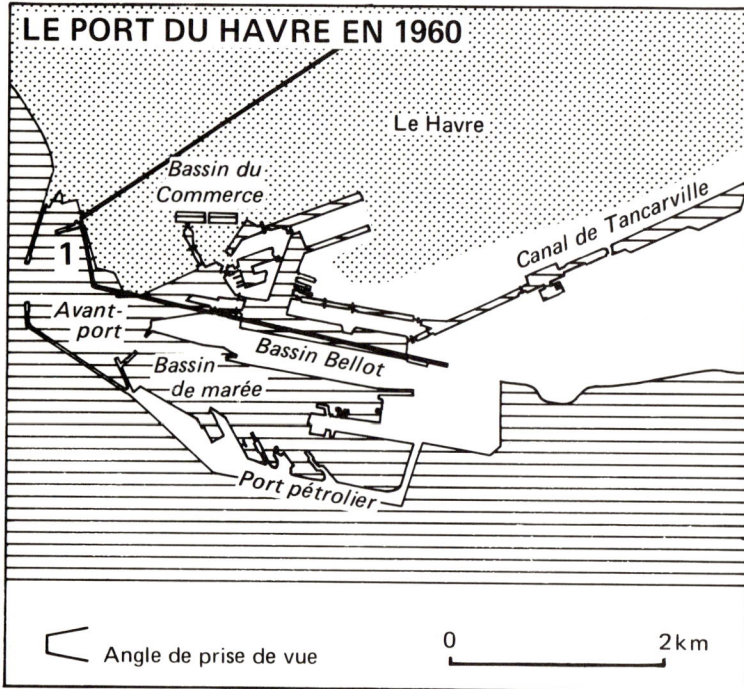

FIGURE 4.14. Le Havre in 1960 ("Angle de prise de vue" refers to a photograph accompanying the map in *La Documentation Photographique*, No. 6028 (1977): *Le Bassin Parisien*; © La Documentation Française.)

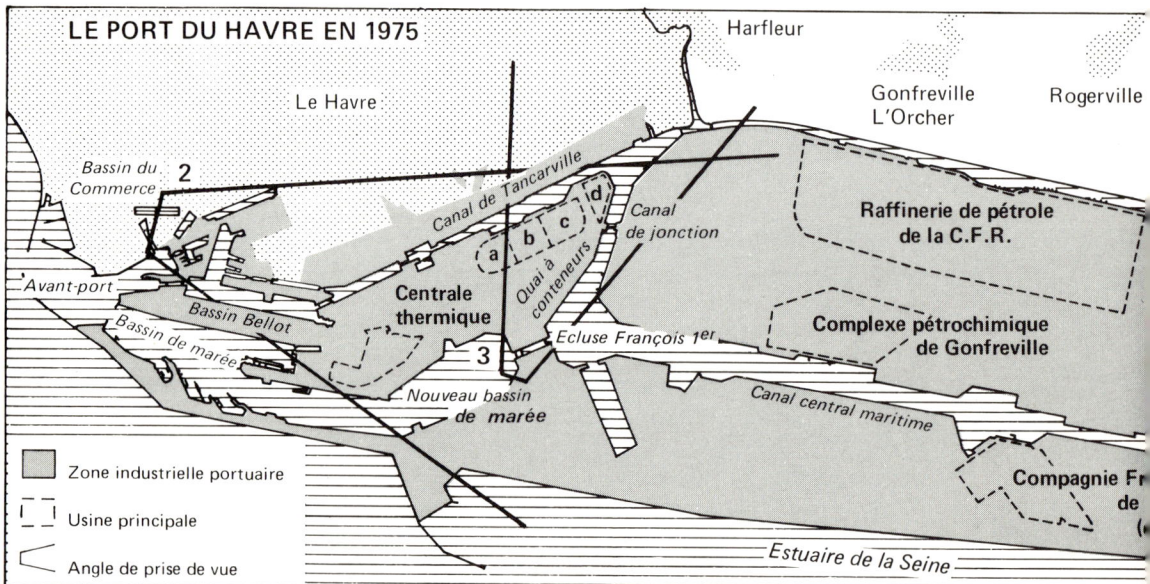

FIGURE 4.15. Le Havre in 1975 ("Angle de prise de vue" refers to a photograph accompanying the ma

Liners such as the *Ile de France, Paris, Champlain* and latterly *France* used to ply the Le Havre–New York route in four days. The passenger trade declined to almost nothing in the second half of the 20th century. But the petroleum traffic which developed rapidly from the 1930s onwards (war years excepted) more than compensated for this, as did also in the 1960s the growth of cross-channel ferry services of the "roll-on, roll-off" variety undertaken by the Townsend Thoresen and Normandy Ferries links between Southampton and Le Havre. Apart from the passenger traffic (both foot and car borne) the ferries have an extensive trade in truck borne loads coming from France and other EEC countries into Britain and vice versa.

Questions

7. Some of the Le Havre original industries were of the warehousing, flour milling, sugar refining, petroleum refining variety. How can you explain this?

More recent industries are shown in Table 5. Classify those industries into 2 groups

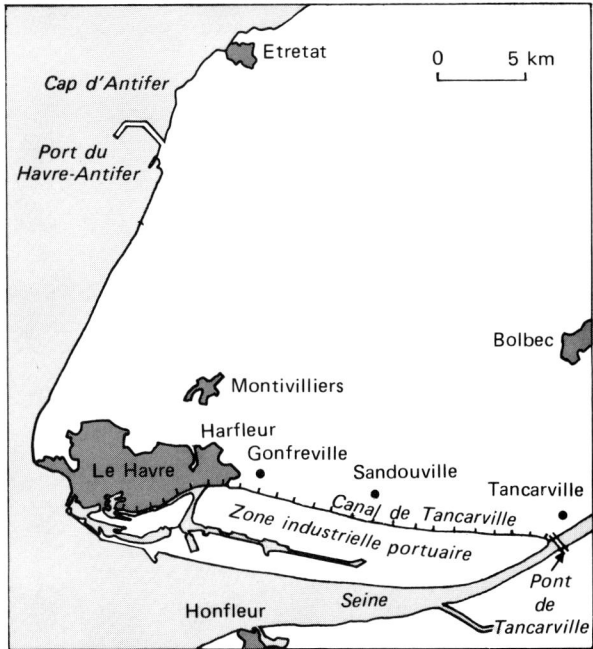

FIGURE 4.16. The outport at Cap d'Antifer in relation to Le Havre (From *La Documentation Photographique* No. 6028 (1977): *Le Bassin Parisien*; © La Documentation Française.)

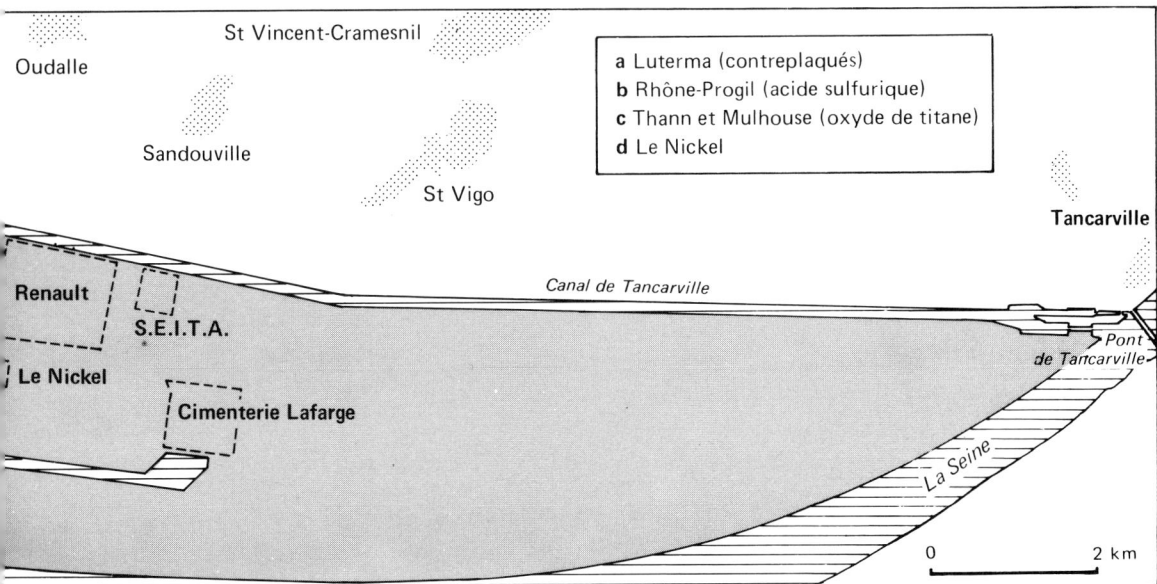

a Luterma (contreplaqués)
b Rhône-Progil (acide sulfurique)
c Thann et Mulhouse (oxyde de titane)
d Le Nickel

Documentation Photographique, No. 6028 (1977): *Le Bassin Parisien*; © La Documentation Française.)

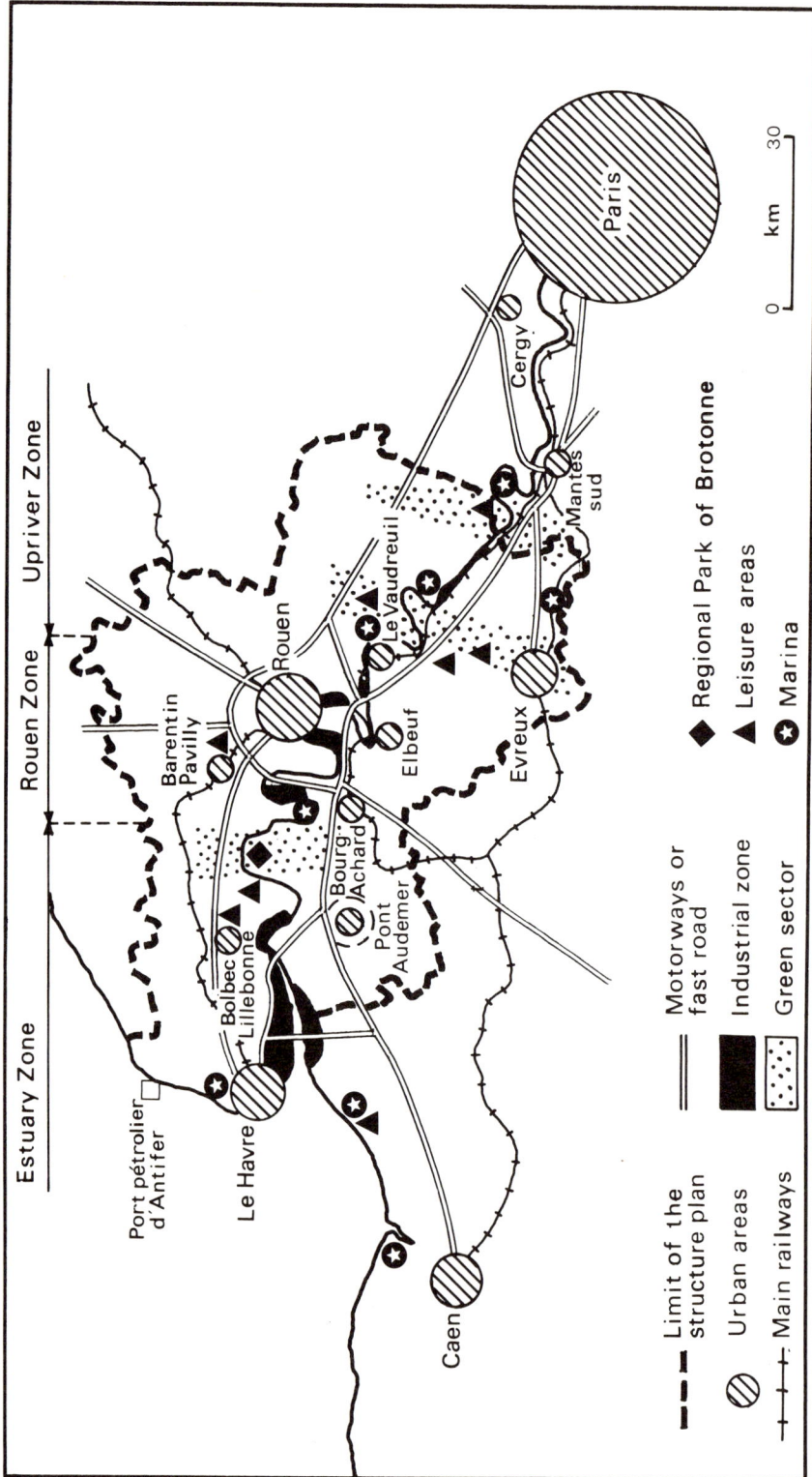

FIGURE 4.17. The Lower Seine Structure Plan

TABLE 5. New enterprises in the region of Le Havre

Companies	Production
Cie Francaise Raffinage (C.F.R.)	Refinery of hydro-carbons —Ethylene —Buthylene —Isobutylene —Propylene Research Centre for radio-active substances
Manolene (Manuf. Normande de polyethylene)	Polythenes
Orogil	Additives for lubricating oils and alkyl-phenol detergent resins
Petrosynthese	Dodecylbenzene "Alkane" Detergents
Luterma	Laminates Plywood
Marles Kuhlmann	Glycol and ethylene oxides
Thann et Mulhouse	Titanium oxides
Saint-Gobain	Sulphuric acids
Goodyear France	Tyres
S.O.M.E.C.	Boiler-works
S.N.E.C.M.A. Normandy	Engineering
Renault Sandouville	Car assembly and body-building

and indicate the basis of your classification. Which group is clearly related to imported raw materials? On Figure 4.15 the cement manufacturing plant (Lafarge Co.) is located at the end of the latest dock basin to be dug on the flood plain of the Seine. What are the advantages of this location from the point of view of (a) the supply of raw materials, (b) the transport of the finished product? (the manufacture of cement requires chalk or limestone, clay and a fuel) (see also Fig. 4.13).

Is the area available likely to limit industrial expansion in the near future? Give reasons for your answer (see Fig. 4.15). Why is the land available for expansion likely to be cheap?

Read

8. From what you have found out, it will be clear that in the case of Le Havre an early decision to develop a port in the area seems to have proved a reasonable one. First, the physical nature of the site enabled and continues to enable an expansion of both the dock basins and associated warehouses and industries. Secondly, the residential areas which accommodate the growing population of the town have been able to develop on the chalk plateau overlooking the lower town and port, the higher class residential suburb of Sainte Adresse lying due north of the "avant port" faces the open sea and contains many attractive villas perched along the land rising up to the plateau. Part of it has been picturesquely labelled the "Nice-Havrais", though inevitably weather-wise the inhabitants do not enjoy Nice sunshine or temperatures.

The nature of the port's trade and the emphasis of its industries has also changed. But the industries illustrate clearly what is general to most ports, namely a warehousing or *entrepôt* function, and a group of industries clearly related to raw material imports, these are generally called *port industries*.

Questions

9. What we have yet to find out is why all this expansion took place.
Explain why you think the growth in petroleum imports occurred.
What has happened to the French Gross Domestic Product since 1950? (See Table 6.)

TABLE 6. French Gross Domestic Product, 1950–70

Year	French Gross Domestic Product (million Francs) (1959 values)
1950	166 000
1960	262 000
1970	405 000

What has been the growth in the population of Le Havre from 1952 to 1979? Since this has not been due only to a national increase, from where has the rest of the population come? What sectors of the economy of France have been able to use less labour to make migration to such expanding areas as Le Havre possible?

Read
10. The economic and physical development of a port like Le Havre is seldom, in a developed country, something which occurs independently of other developments in the region and the country. To complete the study, the map in Figure 4.17 shows the position of Le Havre in the overall structure plan for the development of the lower Seine valley in France.

Question
11. What do you think the planners are trying to do for the lower Seine valley?

This could then lead to a teaching unit on the concept of regional planning within the French context.

An alternative way of programming teaching units is to divide the class being taught into groups and to set each group specific tasks and then to bring the groups together in order to pool their findings. In the example which follows, a group of average ability 4th year (15-year-old) students were asked to find out the environmental conditions under which hydro-electric power was produced in four separate locations with a view to discovering what might be an ideal site. Rather than have the whole class study each example, the class was divided into four groups, each group studying one scheme. In the unit which follows, an attempt was made to find out whether the principles of HEP dam location which had been learned, could be applied in a new hypothetical location. This is a way of continuing the use of case studies with a more general but hypothetical situation.

Hydro-electric Power: Unit I
4th Form (15 years old)

Objectives
To find out the physical conditions of HEP station location in four sample areas in the world, and infer the principles of HEP dam location.

Resources
1. Class to divide into four groups. Each group to investigate one of four different HEP schemes.
   (a) High dam at Aswan, Egypt (Group 1 Questionnaire and Resources).
   (b) The Dordogne scheme in France (Group 2 Questionnaire and Resources).
   (c) The Blaenau Ffestiniog scheme in Wales (Group 3 Questionnaire and Resources).
   (d) The Volta or Akosombo dam in Ghana (Group 4 Questionnaire and Resources).
   A questionnaire is provided to make each student focus on the environmental conditions in each location.
2. When the task is completed bring the class together again, each group to state its findings orally.
3. Class to summarize the conditions found on table provided (Table 7).
4. Conclude on what would appear to be ideal physical conditions for the setting up of an HEP dam and generating station namely:
   (a) A site with strong rock foundations to anchor the dam securely;
   (b) A narrow, deep valley or gorge to store water;
   (c) A steady and plentiful source of water, hence the need for a good catchment area where precipitation is high, therefore often in a highland area.
These conditions to be noted.

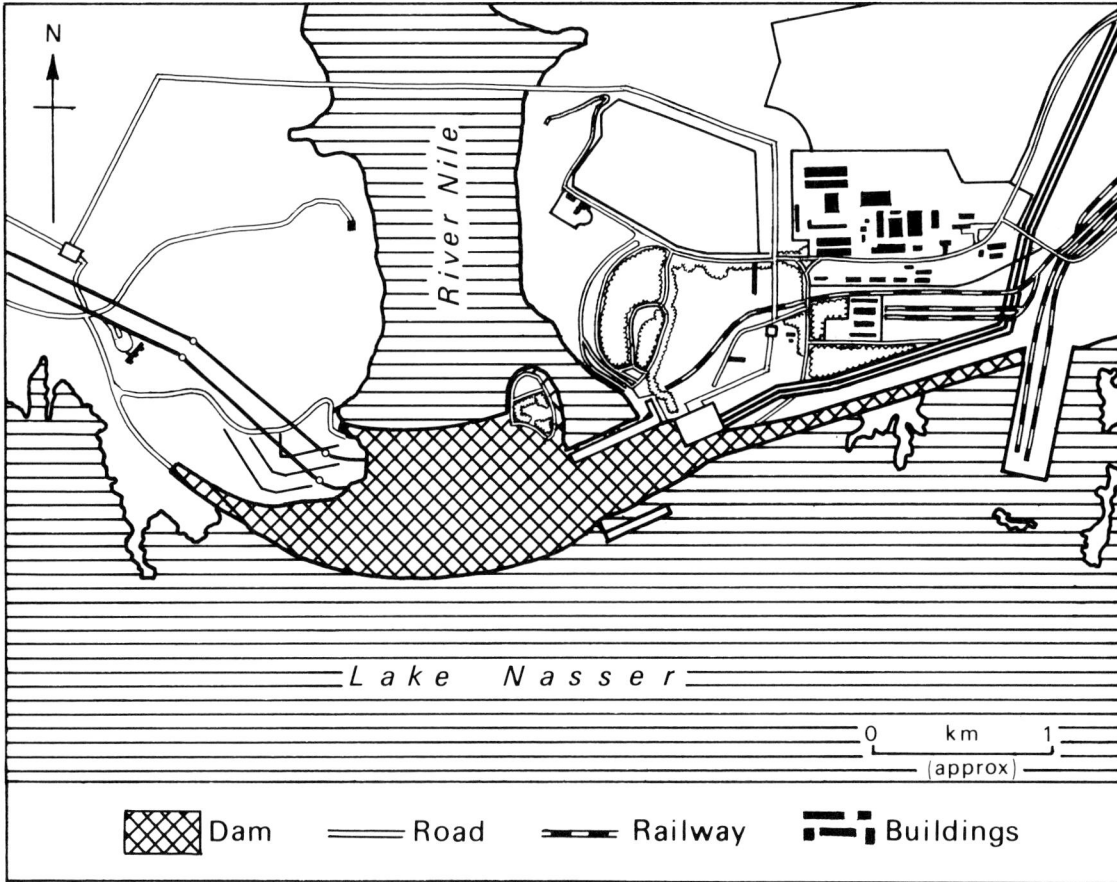

FIGURE 4.18. (a) Plan of the Aswan High Dam

5. Comment on the fact that seldom are all these conditions met. Why then are HEP dams still built?

Group 1: Aswan High Dam

The difficulties which have to be overcome in such a project are considerable. The water engineers who prepare the plans must first calculate how much water there is in the river. They must take account both of the annual flood and of the great variations from year to year. The main purpose of the Aswan High Dam [is] to smooth out these variations. In flood time the reservoir [is] allowed to fill up, its capacity being so great that all the flood water [can] be held back. The old Aswan barrage, which stands four miles downstream of the new site, was designed to trap only a

small fraction of this. . . . The new dam [holds] enough water for a great increase of irrigation and [also provides] storage for a long-term reserve.

All the planning and paper work before the dam [was] built [cost] money, but this is nothing compared with the expenses of the actual construction. Thousands of men, skilled engineers and scientists, lorry and crane drivers, mechanics and labourers [had to] be hired, taken to the site of the dam, fed and housed for up to ten . . . years. Every single machine or tool which they [used] from giant walking excavators to hammers and nails, [had to] be carried there. . . . Limestone and clay [had to] be quarried to make cement, millions of tons of broken rock, sand and gravel [had to] be found, miles of steel rods [had to] be

1 Dune sand
2 Coarse sand
3 Stone with sand
4 Rock
5 Crushed stone
6 Grout curtain prevents seepage below dam
7 Central clay core
8 River bed
9 Reservoir empty
10 Reservoir full
11 Sand gravel and boulders of the Nile
12 Bed rock

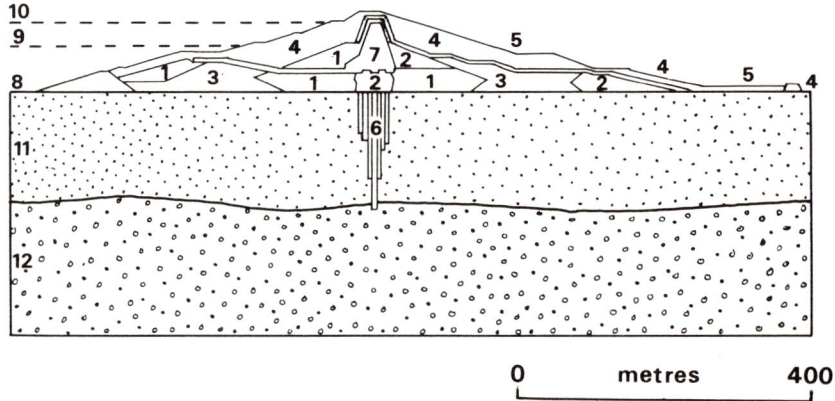

FIGURE 4.18. (b) Cross-section of the Aswan High Dam

brought to strengthen concrete sluice gates. To achieve all this special roads and railways [had] to be built, [as well as] the network of canals to distribute the water.....

The construction site [covered] over 50 kilometres... and [employed] about 10,000 workers and technicians. Three teams [relieved] each other in the task of breaking up stones, digging, levelling and regulating the bed of a new river and of the canal which [diverted] the course of the Nile while work on the dam [was] in progress.... The [completed dam requires] only a small number of engineers... to keep it in repair. Experts in agriculture and government officials . . . control the use of the water and the electricity generated in the power stations associated with both the new and the old dams. . . .

The Aswan site has serious disadvantages. The old Aswan Dam was based on solid granite and was itself built of stone. Under the new dam, however, is a thick bed of sand, gravel and boulders, so that the dam itself [had] to be an extremely broad, thick embankment of clay, with a sand and rock core to give strength (Fig. 4.18). A concrete wall or barrage would not stand firmly on such a foundation. More important, the reservoir behind the dam... extends [480 km] upstream, exposing a vast surface area to the direct rays of the desert sun. It has been calculated that 10% of the water [is] lost by evaporation each year. This represents about ten thousand million

tons, ... about twice the present capacity of the old Aswan Dam. These huge losses cannot be avoided if the reservoir stands in the hottest part of the desert.

It has been found that when liquid *cetyl alcohol* is poured on to the surface of a small lake it spreads out in a layer one molecule thick; the water is protected from the sun and evaporation losses are reduced by a third. Unfortunately, this method is not successful on large bodies of water. Waves, which [are] quite large on a reservoir the size of Aswan, ... break up the film, so that it cannot maintain itself. Even if the method were more successful, the problems of manufacturing enough cetyl alcohol to cover a vast expanse like the [Nasser] reservoir, and replenishing wastage from time to time, would be quite considerable. (Simons, M. *Deserts*. London: O.U.P., 1967, pp. 44–8.)

Questionnaire
1. In what country is the High Dam at Aswan?
....
2. On which river is the dam?
....
3. What is the general level of the land around the dam? (See your atlas and the documents.)
....
4. On what foundations is the dam built? (See Fig. 4.18(b.)
....

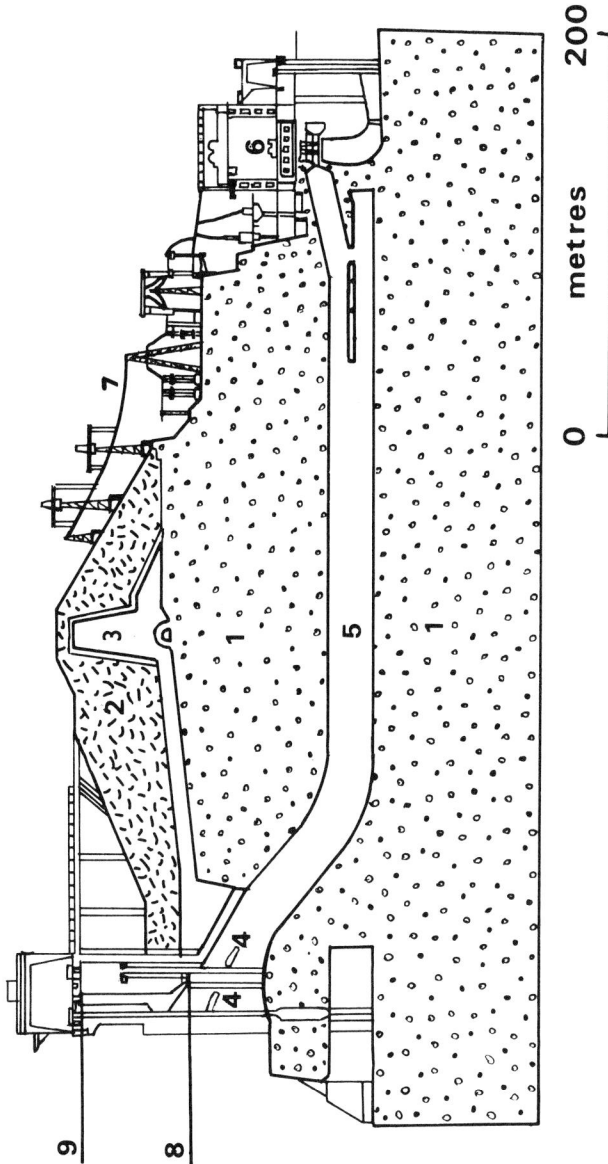

FIGURE 4.18. (c) Spillway structure of the Aswan High Dam

1 Bed rock
2 Dam body sand and broken stone
3 Dam core clay
4 Sluice gates
5 Tunnel through rock below dam
6 Water turbines and generator
7 Power distribution cables
8 Reservoir empty
9 Reservoir full

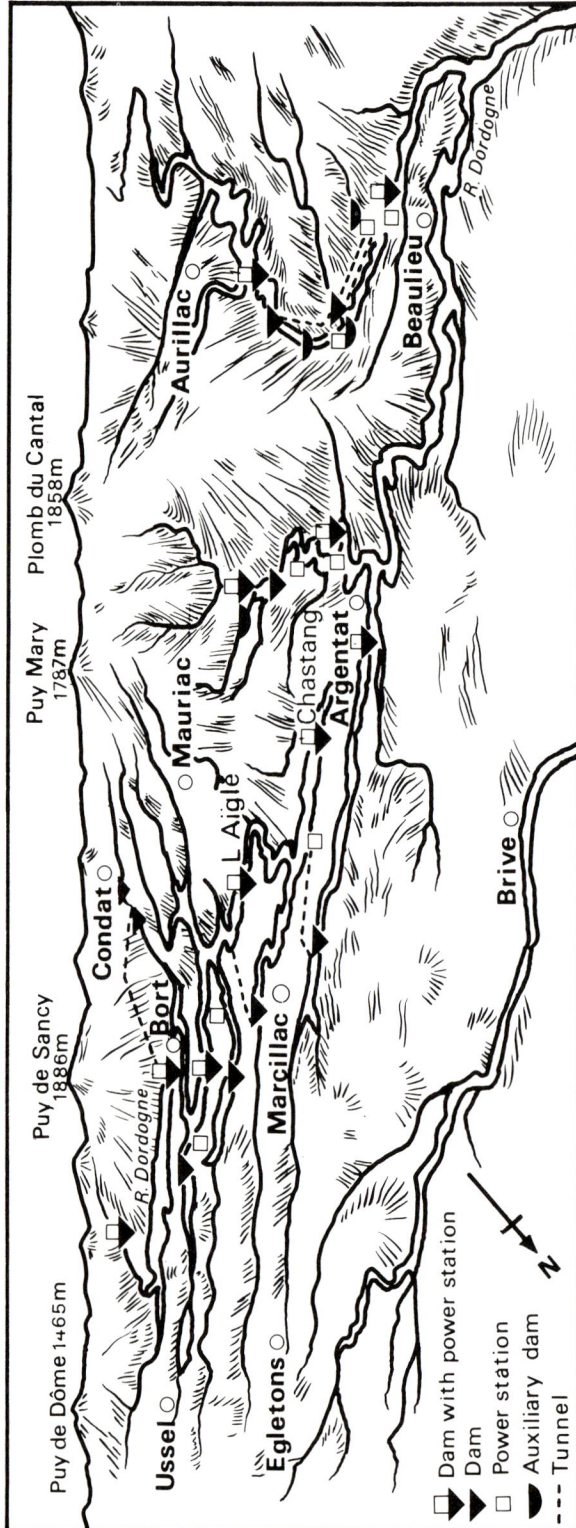

FIGURE 4.19. Pictorial map of the Dordogne power stations

5. Is the river in a narrow valley or gorge near the dam? (See atlas.) What effect does the shape of the valley have on the lake behind the dam?
....
....

6. What is the rainfall like in the area of the dam? (See climate graphs in the atlas.)
....

7. Why do you think the dam was built though it costs so much money?
....

Group 2: HEP in the Dordogne
Questionnaire
1. Where is the river Dordogne?
....

2. What do you notice about the HEP scheme from the pictorial map? (Fig. 4.19)
....
....

3. What can you say about the land around the Dordogne Valley? (See atlas.)
....

PLATE 11   The barrage at l'Aigle on the Dordogne

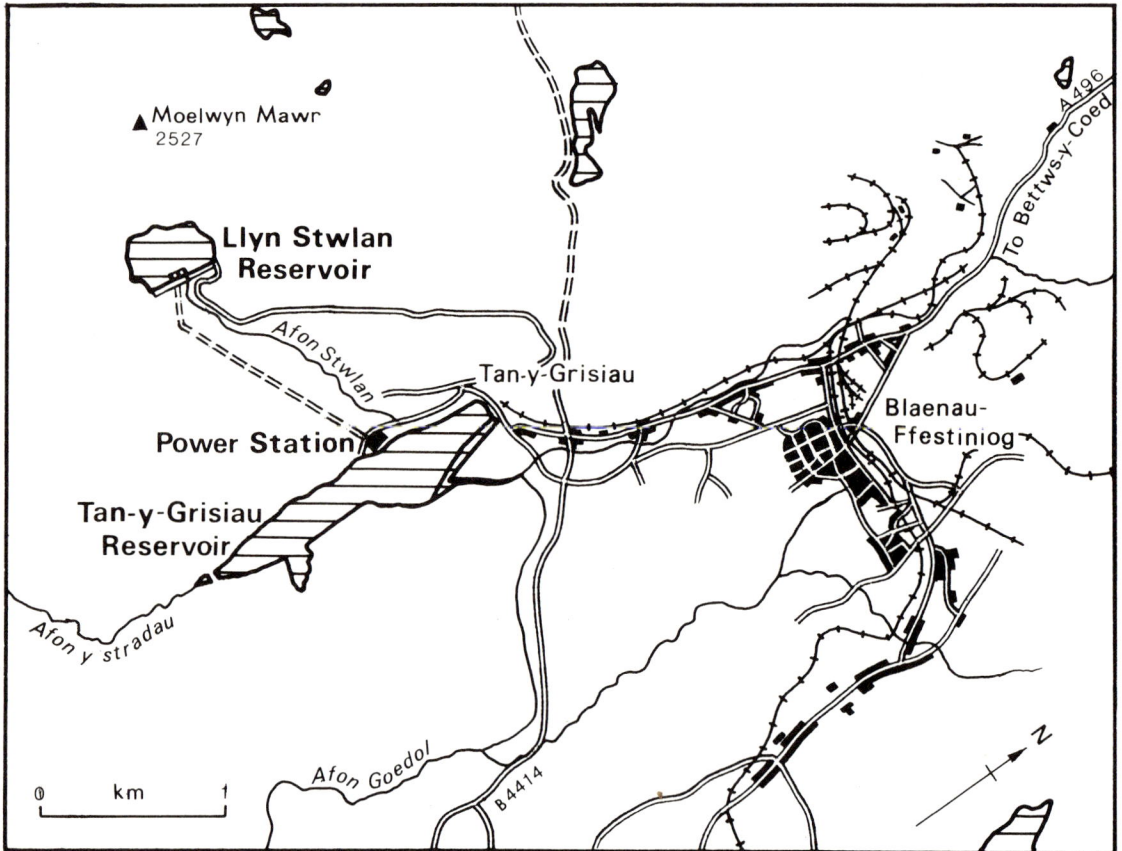

FIGURE 4.20. Map showing Blaenau Ffestiniog HEP scheme. (When few people want electricity, water is pumped back up a pipe to the Llyn Stwlan storage reservoir.)

4. Describe the site of the "barrage de l'Aigle" (Plate 11)
   Is the dam likely to be securely anchored?
   ....
   ....

5. What is the rainfall like in the area of the Dordogne valley?
   ....
   ....

Group 3: Blaenau Ffestiniog HEP Scheme Questionnaire
1. Where is Blaenau Ffestiniog? (See atlas.)
   ....

2. Is the dam on a big river? (Fig. 4.20)
   What sort of a physical feature is Llyn Stwlan reservoir in?
   ....
   ....

3. What is the land like around the dam? (See Plate 12 and atlas.)
   ....

4. What sort of foundations is the dam likely to rest on?
   ....

5. Is the rainfall in the area likely to be adequate? Give reasons for your answer.
   ....
   ....

6. What is a special feature of this HEP scheme?
   ....

Group 4: Volta Power Scheme
   ...The Hydro-electric potential of the River Volta, which had already been assessed by various bodies, was an obvious candidate for investment.... The problem for a Developing Country like Ghana, of course, was how to finance such a venture. After a long period of

PLATE 12   Llyn Stwlan storage reservoir

negotiation spurred on by Nkrumah, the problem was solved by a kind of package deal which encompassed both the dam and the aluminium smelter. . . . An American company—Kaiser Aluminium—agreed to finance the smelter. The company, called VALCO in Ghana, the Volta Aluminium Company, also secured very cheap rates for electricity . . . The building of the smelter assured the economic viability of the dam. . . . The most populous part of the country now has a reliable electricity grid and there is plenty of spare capacity . . . . The Volta River Authority—the body in charge of the whole scheme—and VALCO are staffed right up to top management by Ghanaians. . . .

The Volta gorge at Akosombo is a reasonable site for the dam. The trouble is, the upstream relief is extremely gradual and has caused the lake to spread over a wide area. In some places in fact, the relief is so gradual that vast areas alternate between swamp and dry land as the lake undergoes quite small fluctuations in level. By any reckoning the 3,275 sq. ml. (8,200 sq. kms.) of Lake Volta is a high price to pay for the relatively modest 250 ft. (76 m.) head of water at Akosombo. The lake has changed the human geography of a large slice of Ghana. Fortunately, population densities and soil fertility in the flooded area were low. Nevertheless, 80,000 people have had to be rehoused. People of different customs, different social structures and different languages have been thrown together and sifted around throughout the resettlement areas . . . .

Photograph by Peter Larsen

PLATE 13   The Akosombo Dam on the Volta in a rare period of high water level

...The Ghanaians hope to see transport eventually improved when Lake Volta becomes an important inland waterway. They also have plans to establish refrigeration plants around the shore to handle fish caught in the lake. (BBC. *Living in a Developing Country*, Teachers' Notes, rev. ed., 1975, pp. 20–21.)

Questionnaire

1. On what river is the dam?
   ....

2. What is the name of the dam?
   ....

3. In what country is it situated?
   ....

4. What is the land like around the dam? (See Plate 13 and atlas.)
   ....

5. Why was that particular site chosen? (See notes.) Are the foundations likely to be on rock?
   ....
   ....

6. What is the rainfall like in the area of the dam?
   ....

7. For what is the electricity generated used?
   ....

8. Were there any other benefits in building the dam?
   ....

**Hydro-electric Power: Unit II
4th Form (15 years old)**

Objective

To apply the principles of HEP dam location in a hypothetical situation.

Resources

Map of Fiume river basin (Fig. 4.22)
Table for completion by students

Procedure

1. Indicate purpose of the exercise.
2. Students are to come to a decision as to where to locate a dam on the river Fiume in

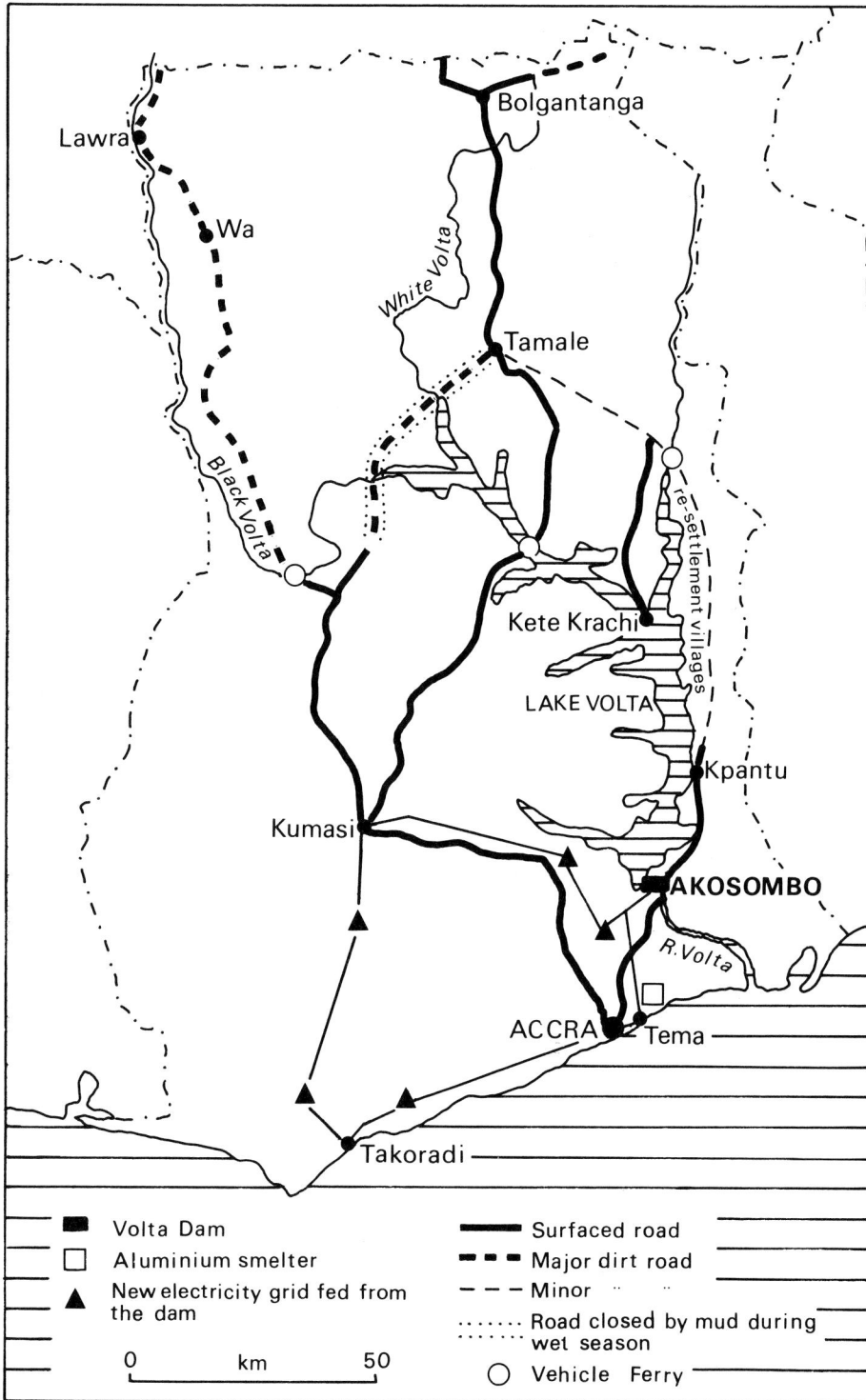

FIGURE 4.21. Map of the Volta HEP scheme

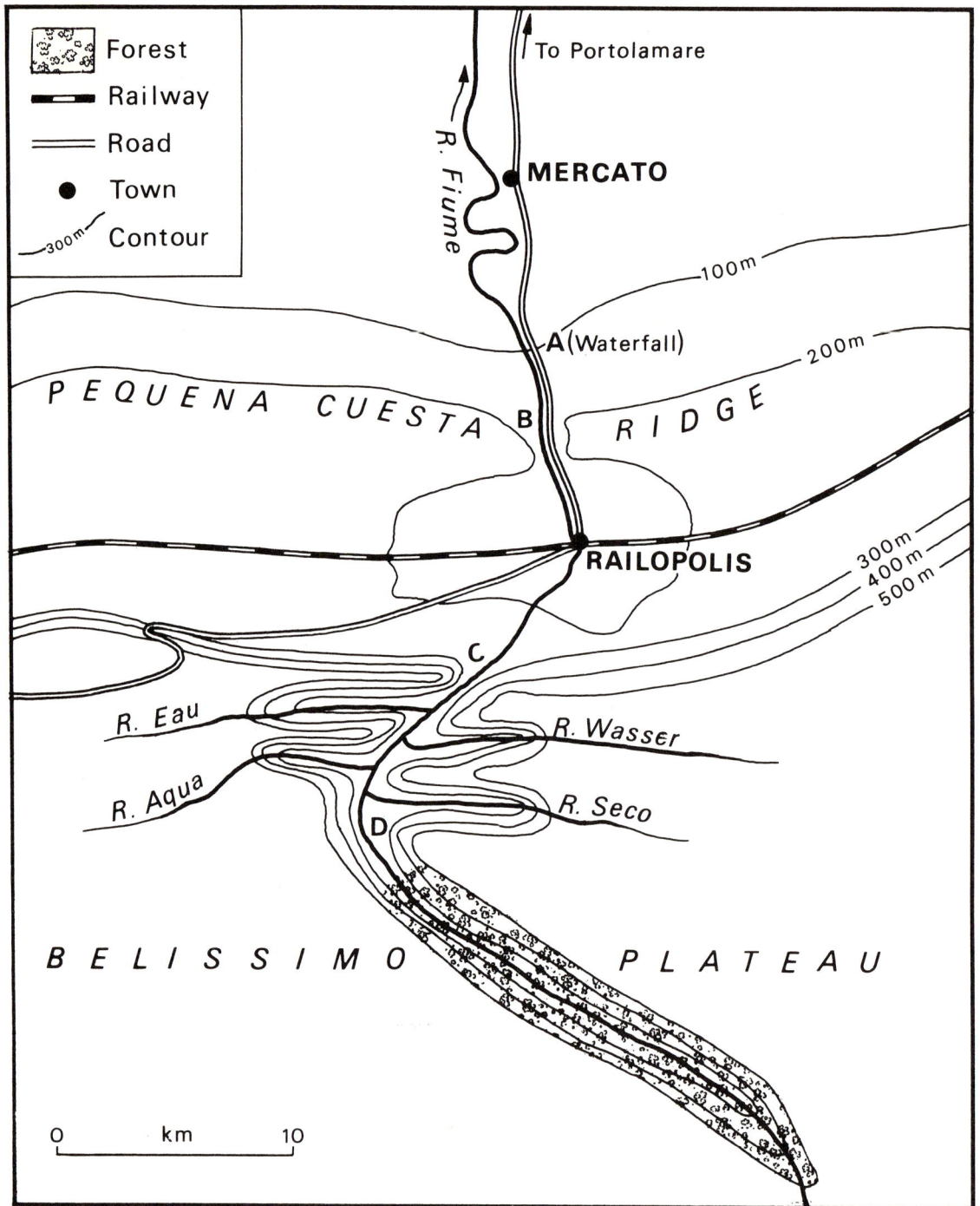

FIGURE 4.22. Possible sites for a dam on the River Fiume in Unaria

TABLE 7. HEP Schemes

Conditions	High Dam, Aswan	Blaenau Ffestiniog Scheme	Akosombo Dam	Dordogne Valley Scheme
Location (Country and area)				
Rainfall				
General height of land				
Shape of valley		•		
Foundations				
Special Features				

the country of Unaria. Four possible sites are indicated A,B,C and D on the map (Fig. 4.22).

3. Students are first to list the advantages and disadvantages of each site on the table provided (Table 8), given the environmental conditions listed at the top of the table.
4. A decision and the reasons for this are to be recorded at the bottom of the table.
5. Discuss with the class the sites which have been chosen and correct any misunderstandings.

As in the case of many teaching units, only one main objective is indicated to obviate the need to enumerate a number of these, but clearly in the above case the unit is also useful in helping to reinforce map reading and in particular the understanding of the landforms associated with certain contour patterns (valley, gorge, scarp, basin, plateau).

4.3 SKILL TEACHING

As indicated at the end of the last teaching unit, the teaching of skills, such as those of contour pattern reading, often occurs incidentally in the process of developing the understanding of a principle or concept. Nevertheless, certain skills are so important in enabling students to develop their abilities further that special consideration must be given to them. The word "skill" has itself been used in a variety of ways, including that of general intellectual skills such as the ability to undertake inductive and deductive reasoning. These are, however, so widespread in the education process that they are implicit in almost all activities. For example, in the two teaching units on the location of a hydro-electric power dam, the first unit implied the development of inductive thinking, arguing from particular cases to general principles; the second unit implied the development of deductive thinking: students were posing the questions: "If

TABLE 8. Choosing a site for a dam on the River
Fiume in Unaria

Information about the area:
1. Most of the land below 300 metres is cul-
   tivated, unless otherwise indicated on the
   map.
2. The climate is hot and dry in summer, warm
   and rainy in winter. There are few frosts.
3. The river occasionally floods near Mercato.
4. The waterfall at A is a tourist attraction.

	Advantages	Disadvantages
Site A		
Site B		
Site C		
Site D		

My choice would be site   because ....

I locate a dam at A (or B or C or D) what will
happen to the water behind the dam? where will it
spread? will there be enough? will it damage
property, drown arable land or forest?" and so
on. Because of the general prevalence of such
intellectual processes, I would prefer to reserve
the term "skill" for psycho-motor skills
associated with maps and diagrams, and the
particular skills implied by certain techniques of
analysis of a mathematical or statistical type.
   Let us take a simple example of the latter.
Suppose that in the context of a study of the
British Isles, a detailed study is made of the
nature and functions of one region like South
Lancashire and that this may be used as the
setting for teaching the concept and technique of
calculating the location quotient. A teaching unit
may be devoted to an analysis of what the
physical and cultural landscapes look like (Unit I
below) and a second unit (Unit II) to explicating
the concept and practising the calculation of the
location quotient for various South Lancashire
industries. In both units the main resource used
is a film strip on South Lancashire which
contains evidence of the landscapes and of the
employment structure in the area. The frames of
the film strip used are not reproduced here for
reasons of economy, but it is one which is easily
available and many schools may possess a copy.
What needs to be made clear at the outset is that
the teaching of skills necessitates an element of
expository teaching. Psycho-motor skills and
techniques of analysis can most economically be
taught by demonstration and practice rather than
by guided discovery.

South Lancashire: Unit I
5th Form (16 years old)

Objective
   To describe and contrast the rural and urban
landscapes of South Lancashire.

Resources
   Atlas maps
   South Lancashire filmstrip (Common
      Ground)
   Filmstrip projector

Procedure
   Oral work
1. Locate South Lancashire area in Atlas.
   (Distance from London, position in
   England.)
2. What can we find out about South
   Lancashire from the Atlas map?
   (Elementary facts about relief, distribution
   of towns and some transport routes. Locate
   these.)
3. What further evidence required? (Visual
   evidence, descriptions.)
4. Show:
   (a) Frame 2—Winter Hill——obtain
       description of relief and land use and
       locate.

(b) Frame 3—Altcar—obtain description of relief and land use and locate.

(c) Frame 4—Upholland area—obtain description of relief and land use and locate.

(d) Frame 7—Dune coast—obtain description of relief and land use and locate.

Explain in terms of geological map (on filmstrip).

Written work

5. Main rural landscapes of South Lancashire (Brief notes on each).

Oral work

6. Show:
   (a) Frame 11—Heywood spinning works—describe.
   (b) Frame 15—Old Town, Blackburn.
   (c) Frame 23—New industrial estate.
   (d) Frame 35—New Town, Kirkby.
   Obtain contrast between (a) and (b) and (c) and (d).

Written work

7. Describe and explain contrasts in urban landscapes.

8. Conclude on planned nature of new urban landscape.

South Lancashire: Unit II
5th Form (16 years old)

Objective

To show how the degree of localization of industries in an area may be measured objectively.

Resources

Atlases
Filmstrip projector
South Lancashire filmstrip (Common Ground)
Table of statistics on % employment in South Lancashire and the UK (Table 9)

Procedure

Oral work

1. Recall nature of South Lancashire landscape.

2. Discuss view that such knowledge is insufficient for planning development since it gives a partly impressionistic view of the area and there is a need for accurate statistics. But statistics of what? Let class suggest: employment, output, land use etc.

3. Obtain source of employment statistics and land use information. (Dept of Employment, 2nd L.U. survey maps, local planning office.)

TABLE 9. Occupational Groups in South Lancashire

(Not all industrial groups included)	*UK % Employed*	*South Lancashire % Employed*	*Location quotient*
Agriculture, Forestry & Fishing	1·7	0·4	
Mining and allied industries	1·5	2·2	
Chemicals	1·9	3·6	
Metals and engineering	10·3	16·6	
Textiles	2·3	9·2	
Clothing and footwear	1·6	2·4	
Food, drink and tobacco	3·1	4·2	
Paper and printing	2·4	2·9	
Service industries	50·5	42·8	

$$\text{Location quotient} = \frac{\% \text{ Employed in a given industry in S. Lancs.}}{\% \text{ Employed in a given industry in UK}}$$

4. Discuss method of judging the importance of an industry in South Lancashire. (See graphs from filmstrip.)
   (a) Absolute figures of employment.
   (b) Percentage figures of employment—what do figures show?

Written work
5. Note main occupation groups in north-west England.

Oral work
6. Would such figures give the degree to which an industry was localized in the area concerned?
7. Obtain need for comparison between position of South Lancashire and UK as a whole by using table of comparative percentages (Table 9).

Written work
8. Class to work out location quotient based on labour statistics for South Lancashire.
9. Class to describe what they have found out about:
   (a) the relative importance of various occupations in Lancashire;
   (b) the degree to which industries are localized.
10. Conclude on limitations of method used.

4.3.1 The Problem of Quantitative Methods

Practising the calculation of the location quotient is not just practice in arithmetic, it is a way in which the concept itself may become clear. Since the calculation based on percentages is a simple one it presents no great problems, though if students are just given access to the original data, namely the number of people employed in a given industry, then calculating the percentage employed in each way becomes tedious. Similarly, any statistical device like the calculation of the variance (or Standard Deviation) or one of the Coefficients of Correlation will often require the use of pocket electronic calculators or if the school or college has one, a mini-computer, micro-processor or access to a computer through a terminal. What is important in the teaching strategy is the stressing of the conceptual basis of the coefficient or index or whatever name is given to the technique used and not its detailed calculation. At present it is unlikely that students below Advanced Level Geography will need to calculate such statistics as the Spearman Rank Correlation Coefficient or Kendall's Taw, but even at A-Level it is much more valuable for the student to understand what the coefficient measures and its limitations than to spend hours laboriously calculating such a coefficient.

With the advent of micro-processors and programmable hand calculators, the actual chore of calculation may be left to one side. Unfortunately, students often assume that once a figure has been obtained, their troubles are over, when in fact they have simply been given a piece of evidence which they must now begin to interpret. For example, a calculation of a chi-squared statistic may indicate that the distribution of a random sample of points showing that there is proportionally more arable land on the Yorkshire Wolds than in the Vale of Pickering, is not a result which could have been obtained by the chance distribution of sample points; in the technical language of statisticians the chi-squared statistic is significant. This procedure was made necessary by the fact that a sample of points was selected. Clearly, had the whole of the two areas been surveyed, then there would have been no need to use a chi-squared test. But the results of the test tell us nothing about why there is proportionally more arable land on the Yorkshire Wolds. This requires the setting up of one or more hypotheses about the relationship between arable land use and the perception of physical and economic factors by farmers. In this case it appeared that land drainage was an important physical consideration in determining the differential land use between the two areas. Teachers wishing to develop their curricula in the direction of various statistical and mathematical techniques are referred to standard works in this area which are able to treat the theme at greater length.[31] Many books now exist at A-Level which contain guidance and exercises on the use of such techniques. Essentially, the geography teacher needs to be satisfied that (a) the problem which he is investigating may benefit from quantitative analysis, (b) that data is available to be so analysed and that (c) the techniques at his disposal applied to the data he has will yield useful and valid information. There is little point in doing calculations which yield unusable evidence. In practice, most such techniques may be classified into:

   (a) (i) Mathematical formulae which model a process occurring in the field—as for example in hydrology.

        (ii) More simple indices which measure certain properties, such as a connectivity index in network analysis.

   (b) Statistical techniques whose purposes are one of the following:

        (i) The description of data—e.g. the description of a distribution as in the case of the Standard Deviation.

        (ii) The estimation of a correspondence between the two distributions as in the case of the correlation coefficients.

        (iii) The estimation of significance, namely indicating whether certain distribution patterns could have been obtained by chance or not—this is made necessary by the fact that most statistical observations are carried out on a sample rather than on the whole population. The chi-squared statistic mentioned earlier is a case in point.

Examination board syllabuses will generally indicate with which particular techniques the students will need to be familiar.

4.3.2 Cartographic Skills

Cartographic skills including both map-reading and map-drawing skills need to be developed steadily and consistently throughout the secondary school years. There now exists a substantial, if to some extent conflicting, literature on children's abilities to read maps and to represent spatial relations in map form.[32] This may be referred to for more detail, but although ideally children arriving in secondary schools should be able to read large-scale plans and have some acquaintance with medium-scale (1 : 25 000, 1 : 50 000) maps, in fact owing to the less structured primary school curriculum, a minority of children are able to do so. Inevitably, a number of children will have come across maps in the course of their normal out-of-school experience, so that the teacher will seldom be faced with children for whom a map is a completely unknown quantity. Essentially, maps provide evidence of spatial relationships and may be looked upon as part of the evidence if the problem is one of finding a location or simply working out distances and orientation. More often, the map needs to be used with other evidence. Thus teaching map reading may be done through the posing of simple and then more complex geographical problems in which the map is an important part of the evidence. Thus, in the early years, large-scale plans may be used as part of simple recording exercises in the locality of the school in which particular features are marked on copies of such plans. Medium-scale maps may be used in the process of studying transport networks, or in studying the distribution of woodland in relation to relief; large-scale (1 : 10 000) maps of urban areas may be used in the study of different urban forms and street patterns. In this way, little by little, the pupil and student learns the language of various types of maps, and learns to find out what maps can and cannot tell him. In general, it is best to begin with large-scale maps and gradually introduce smaller-scale maps. However, this cannot be an absolute rule since there may be sound reasons for using small-scale maps to cover large areas. An atlas map is, after all, a very small-scale map and yet if one is to refer to nationwide and worldwide phenomena then such maps need to be used. What is important is that students should realize as soon as possible that such small-scale maps cannot give detailed information on such aspects as town sites or the precise nature of land forms or local climate on climatic maps.

Any particular skill will need to be taught in a gradual way, for example, the drawing of cross-sections to show variations in relief along a given line needs to be practised with simple contour patterns at first and then with slightly more complex patterns and so on. The same would be true of such a skill as that of sketching out the physical divisions of an area shown on a given map or map extract.

The example which follows is of the use of a Dutch 1 : 50 000 map extract in order to present evidence of changes in the planning of settlements in the new polders of Eastern Flevoland compared with the older north-east polders. This is a slightly modified version of the teaching unit which appeared in *Geography In Secondary Education*. The teaching unit forms part of a theme on settlement geography with special reference to the Netherlands. The need

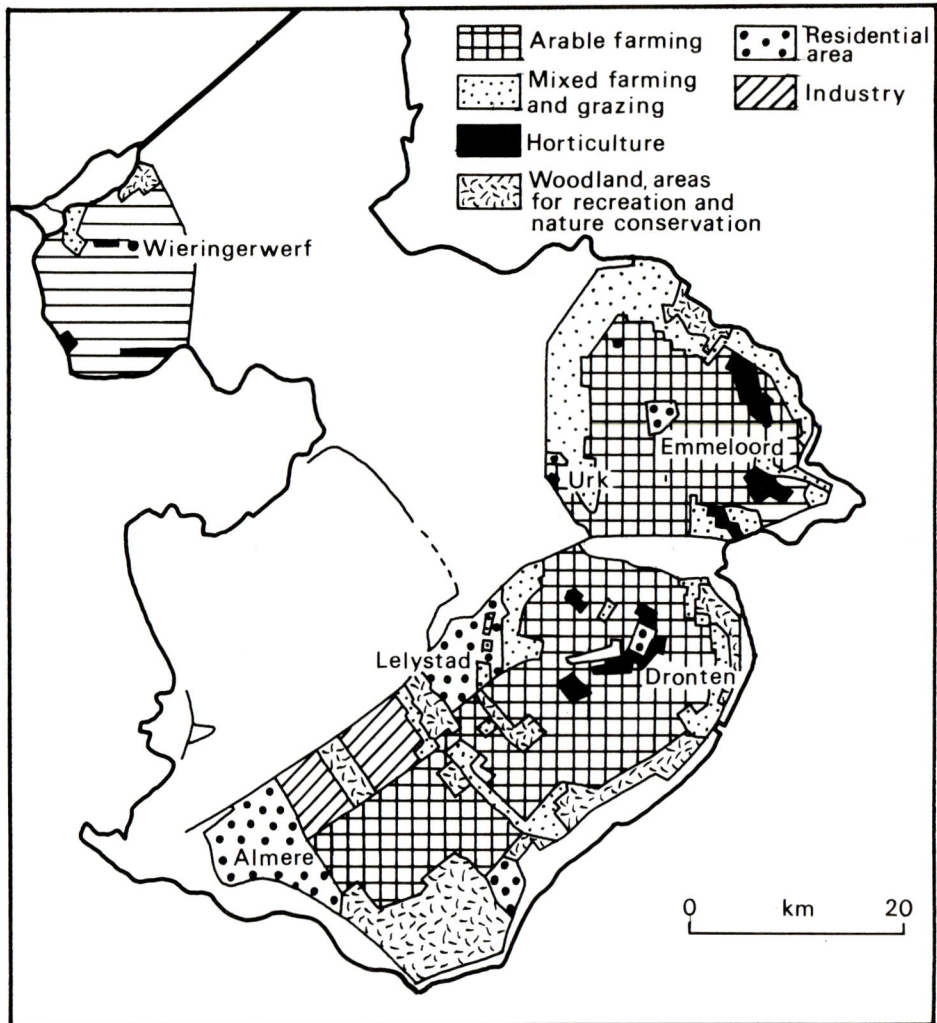

FIGURE 4.23. Land use in the Lake Ijsell polders

for more land and the reclamation of land in the Netherlands was covered in previous units.

The Netherlands Polders
Fifth Form (16 years old)

Objective

To show how changes in economic and technical conditions are resulting in the landscape of new polders being different from that of older polders.

Resources

One set of 1 : 50 000 maps of Zwolle—21 west (Fig. 4.24)

One map of Ijssel Meer area (Fig. 4.23)

One sheet 1 : 100 000 map of Harderwijk (not shown)

Tables of statistics (from I.D.G. *Compact Geography of the Netherlands*, Utrecht, 1974, and I.D.G. *Zuyder Zee/Lake Ijssel Guide,* Utrecht, 1976)

Procedure

Oral discussion

1. Recall nature and purpose of land reclamation in the Netherlands.
2. Establish position of area shown on Zwolle sheet in the Netherlands (Atlas).
3. Obtain possible division of area on map

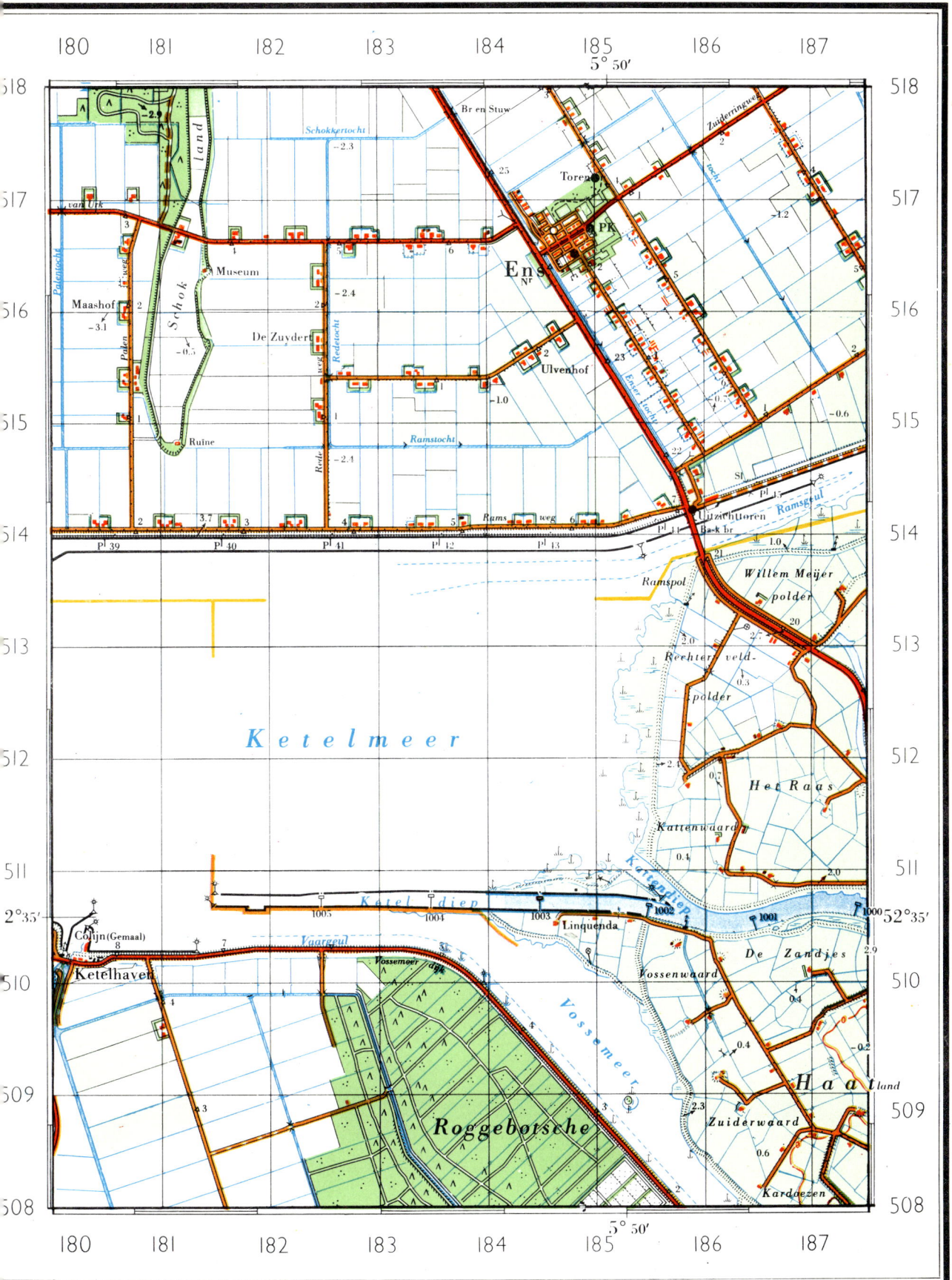

FIGURE 4.24. Reproduction of part of 1 : 50 000 Zwolle sheet—Netherlands

TABLE 10. Planning proposals for North-East Polder and Eastern Flevoland

	N.E. Polder	Planned for E. Flevoland
Size of farm holdings	25 hectares	40 hectares
Population of villages	800 people (500)	1200–1600 people (2500)
Regional town	(Emmeloord)	(Dronten)
	7000 people (15 000)	10 000 people (12 600)
Regional capital		(Lelystad)
for all the polders		100 000 people (15 000)

Figures in brackets indicate actual population

into three parts and the reasons for this division (N.E. polder (to the North), Kampen (to the East), Eastern Flevoland (to the South)).

4. Obtain that Kampen area is different from the others, and why.
5. Study characteristic features of N.E. polder.
   (a) relief
   (b) road network
   (c) settlement
   (d) canals
6. Study land-use map and obtain that most land is arable. (See Fig. 4.23 "Land use in the Lake Ijssel polders", from I.D.G. *Zuyder Zee/Lake Ijssel Guide*, Utrecht, 1976).

Written work
7. Students to write a brief statement under the following headings:
   "N.E. Polder Areas: Landscape of the older Ijsselmeer Polders" (relief, roads, settlements, canals and land use).

Oral discussion
8. Compare Eastern Flevoland with N.E. Polder by examining two 5-square-kilometre samples and conclude that density of farm settlement is lower in spite of similar relief conditions.
9. Examine reasons for this by:
   (a) Looking at 1958 map of Eastern Flevoland (not included here) and noting that land reclamation was only beginning then and that settlement had not begun.
   (b) Studying planning proposals in Table 10 which suggest it is a deliberate act

of policy to reduce the rural population but increase the urban population in the New Polders. (In practice, the villages in the Older Polders are smaller than planned, the urban areas larger than planned.)
   (c) Noting that this reflects a second trend in the Dutch economy—only 7% of the working population is in agriculture and 81% live in towns of more than 10 000 people—therefore planning has to take this into consideration.
   (d) Obtaining that larger holdings are made possible by mechanization.
10. Obtain that density of villages and towns is lower on Eastern Flevoland than in the N.E. Polder and seek reasons for this in:
    (a) Villages with limited functions are not required given improvement in private transport and people's demands for better shopping facilities.
    (b) The economies of scale achieved by having larger schools, etc., given the peculiar religious and political conditions in the Netherlands.
11. Note under the heading, "The landscape of the new polders: Eastern Flevoland":
    (a) differences between newer and older polders;
    (b) reasons for these differences.

Map drawing and other graphical skills are perhaps less emphasized in geographical education than they once were. Nevertheless, there is still some educational value in students learning to co-ordinate hand and eye with a view to drawing sketch maps, charts, graphs and sketches of landscapes. Let me try and justify this statement. First, drawing sketch maps and landscape sketches, given the proper instructions, teaches pupils to concentrate their observations on what is deemed significant. It is a way of selecting graphically the relevant from the irrelevant. The nature of the relevance has to be decided according to the task in hand. Secondly, learning to draw graphs, bar-graphs, divided circles and proportional divided circles helps to reinforce the notion of precision and accuracy where these are important and yet also teaches the need for rounding off figures where accuracy to several decimal points could not be represented graphically. Thirdly, many pupils take pride in producing a visually pleasing map, graph or drawing.

It may also be useful to involve students in transforming data into map form, such as making a map of the countries of the EEC, in which the areas of each country are proportional to population or gross domestic product and so on.

4.4 PROBLEM-SOLVING TEACHING STRATEGIES

One does not want to give students the impression that life, and school life in particular, is a series of problems to which they must find a solution. Nevertheless, the extent to which geographical education has been successful may be judged partly by the enjoyment that students express in travel and the curiosity they manifest in geographical phenomena, but also in the extent to which they can solve what appear to be spatial problems of one sort or another. An example was given earlier in the case of the dam location problem (pp. 48–51). Problem solving may be undertaken at several levels of difficulty. The simplest would involve the application of a concept or skill recently acquired in a new situation. Slightly more difficult would be situations in which the student had to decide what principle or skills were appropriate to the solution of a problem, but with the scope of the problem still limited. For example, the problem of organizing a commercial traveller's round and headquarters in an area with a certain road network, so that he can visit the maximum

number of shops in a given time (see *Teaching Geography Occasional Papers,* No. 9). Still more difficult would be a situation in which not only would the principles or skills to be used have to be decided, but the problem is a multi-factor one with perhaps several stages in its solution. Many industrial location problems are of this type since (1) one has to decide whether to apply a least-cost location principle or one based on profit maximization or a welfare approach; (2) the number of variables to consider may be very large if there are many raw materials, if several types of labour are required, if the market centres are many, if there are environmental and social costs to be considered and so on, and (3) the solution finally accepted may depend on working out the consequences of other solutions, hence the multi-stage apppproach. Some problems require a technical solution only, namely what would be technically feasible in the circumstances, as in the case of the extension of the dock basins in Le Havre; the flood plain but not the chalk plateau may be used as locations. Most are both technical and economic; for example, should the newest dock basins be tidal or isolated from the tides by lock gates? To some extent this is an economic question as well as a technical one, since a tidal dock, if technically possible, would be more expensive to excavate than an isolated basin, and therefore the question becomes: would income from extra ship traffic compensate for the capital expended?

Many real-life problems are complex and it is clear that part of the teacher's task in the planning of teaching units lies in the modification of such problems so that they may be tackled by students at a level which will stretch them without discouraging them. Thus inevitably problem-solving teaching strategies tend to depend on situations which are artificial to some extent, in which the complexities of a real-life problem have been simplified. Indeed, it is the essence of model building to make manageable the inordinate complexities of "reality".

Let us take a simple problem of house location and orientation, which though simplified in detail is based on a real case. Mme Dupont has been left by her father a piece of land in the Mediterranean area of Southern France which is just large enough, according to the planning regulations, to enable her to build a small house on it, to which she hopes to retire when she stops

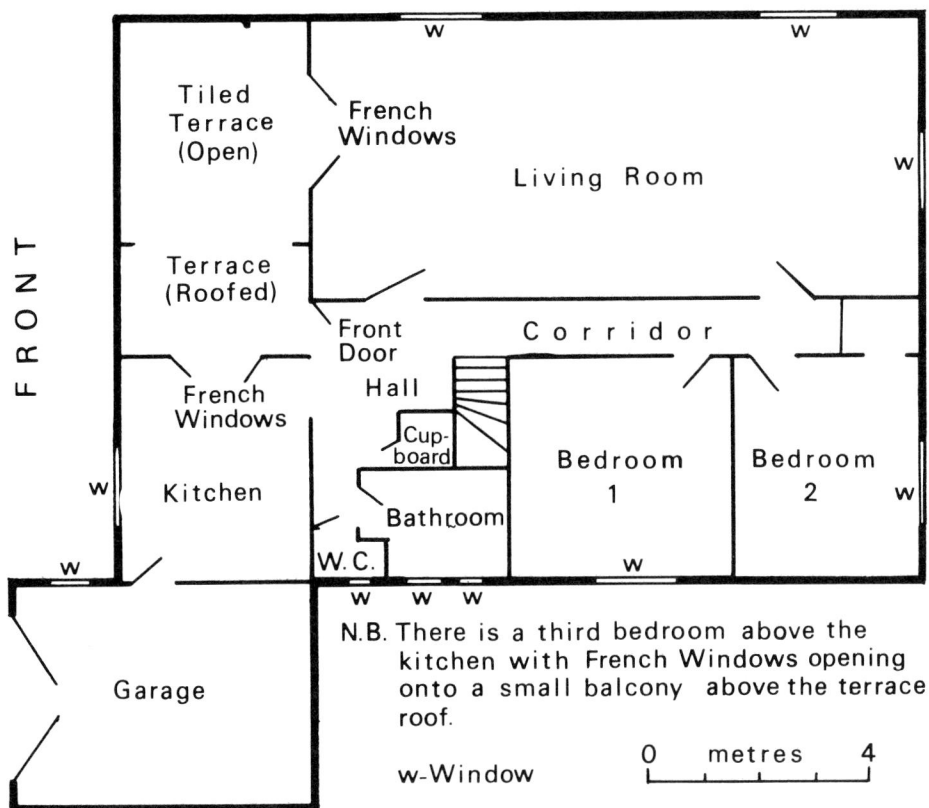

FIGURE 4.25. Plan of Mme Dupont's house

work at 60 years of age. She has a moderate income and cannot afford a specially architect-designed house, but decides to choose a standard design which is illustrated in Figure 4.25. The piece of land she has inherited is a regular rectangular shape and is shown on Figure 4.26 as well as the surrounding plots, some of which have houses on them, some without. The location of the plot in the district is shown in Figure 4.27. The problem is to decide where she will place her house on the plot and with what orientation. Mme Dupont would like to have good views of the limestone hills, the tops of which are only 4 km away, and to benefit from the sun in as many rooms as possible, to keep her heating bills low in winter. She has access to running water from the local town's water supply. Students are asked to justify the location and orientation they have chosen, and to draw it on a copy of Figure 4.26 (the area of the plot). It is important to realize that, though in this case it is possible to refer to

the actual emplacement chosen by Mme Dupont (she placed it in the north-west corner of the plot, with the front facing east), there is no one right answer, even in this case. Someone with different requirements from those of Mme Dupont might have chosen a different emplacement; for example, someone who would have wanted to minimize the cost of access to water and sewerage from the public road.

It would be possible to extend such an exercise to other aspects of geography. For example, the plot of land is substantial by suburban standards, 2160 m², which is over a fifth of an hectare. Few suburban gardens exceed 400 m². The problem is how is she going to lay it out and with what plants? Mme Dupont comes from the western suburbs of Paris where gardens are either vegetable plots or formal gardens of flower beds, lawns and paths, and where the climate and soil are very different. Her plot of land lies on periglacial deposits derived from the limestone

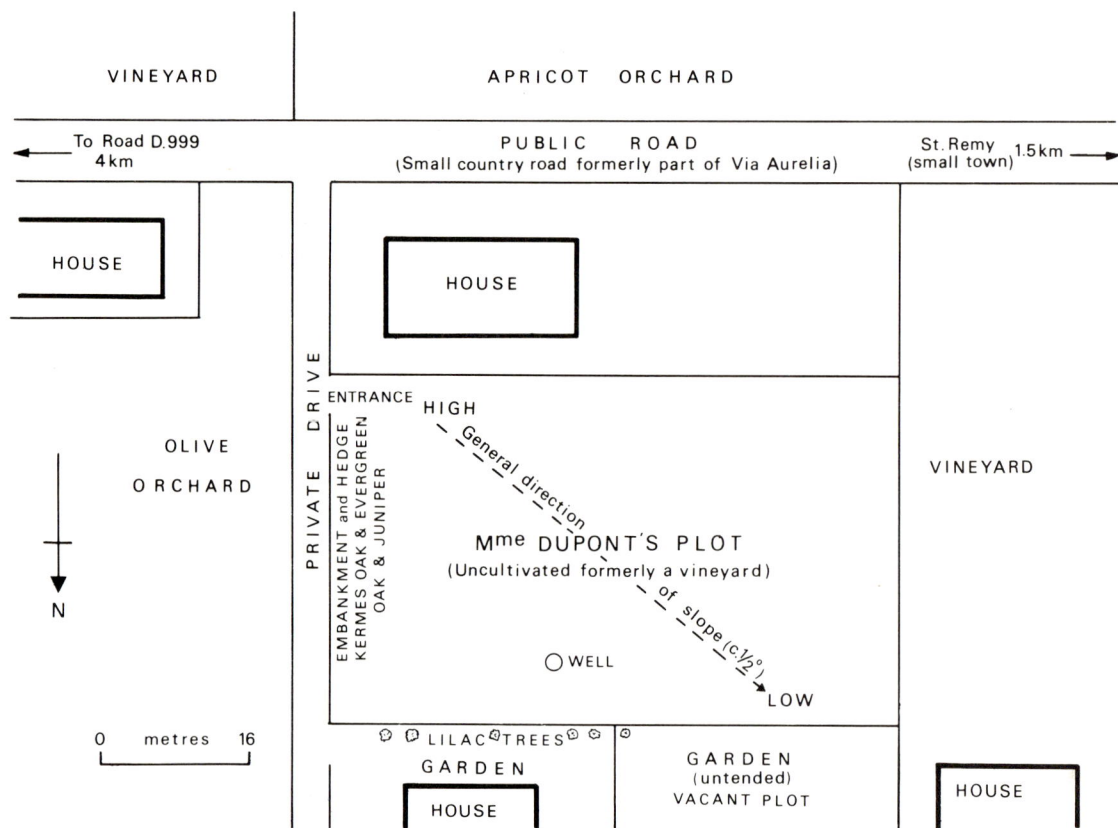

FIGURE 4.26. Mme Dupont's plot of land and its surroundings

hills. The soil is shallow, stony, with a high lime content and extremely pervious. The water table is 14 metres below the surface. She has a deep well dug and an electric pump installed which allows her to pump up about 1000 litres of water per day, but no more. She will be living, when she retires, in an area subject to considerable periods of drought. From May to early September it seldom rains except for the occasional thunderstorm, and with daytime temperatures in the 30°s Celsius evaporation is high. In the autumn and early spring, periods of cyclonic rain occur, but not as frequently as in the Paris area, and evaporation is still an important consideration. Winter is less likely to experience rain than spring or autumn, but when the *Mistral* (a north wind) blows it can be very cold with temperatures from 0° to −8° Celsius at night. Luckily the *Mistral* only blows occasionally, but even without the wind, night frosts

in late December through to late February are not unusual. She could, as some "immigrants" have done, lay out a garden in the style familiar in northern France, but this would probably be unwise—what options are available to her? She could plant Aleppo pines, rosemary, broom, lavender, almond trees, olive trees, pubescent oak, evergreen oak, oleanders, cypress, all of which would resist the drought conditions of the summer. Since she has some water from the well, she could also plant some trees, flowers and possibly vegetables which would require regular watering. Since both she and her husband will not be in their first youth, they will want to lay out a garden that limits the amount of work required. Let us assume Mme Dupont decides that she wants a clump of Aleppo pine for shade in summer, an area of lavender, a drive to the house lined with bushes, a splash of yellow afforded by the broom in bloom, a means

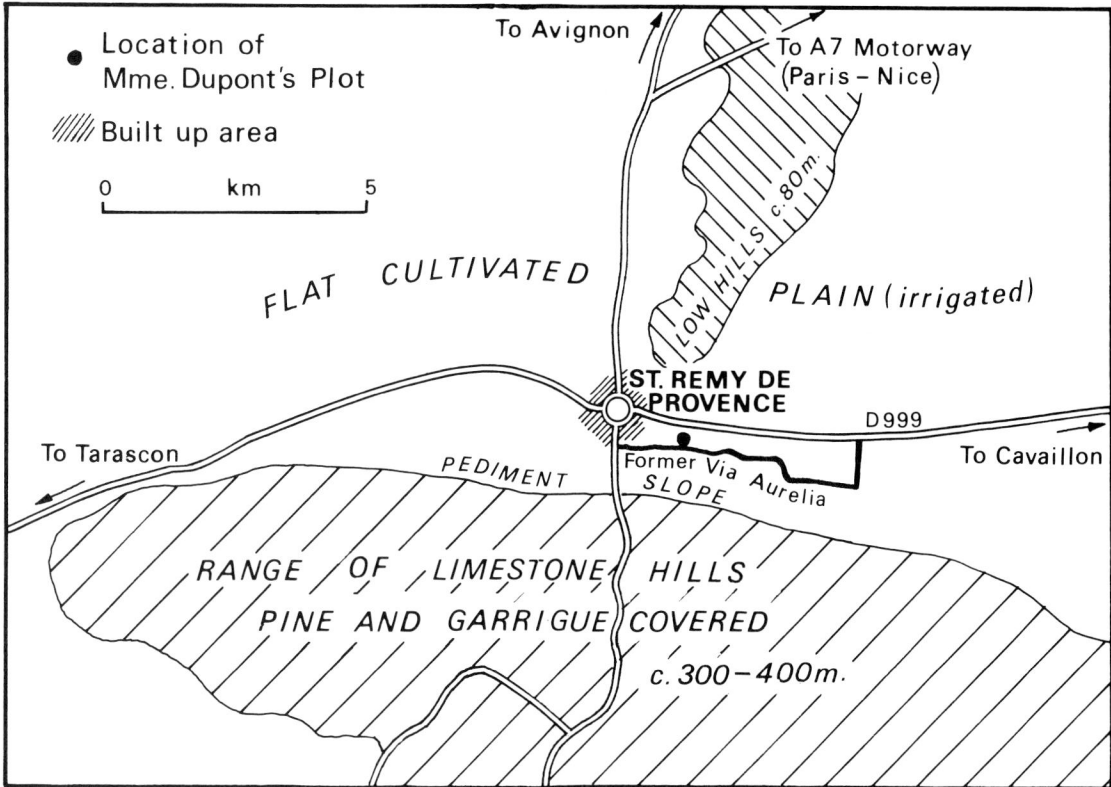

FIGURE 4.27. Location of Mme Dupont's plot of land

of protection from the *Mistral* and a small vegetable garden to grow tomatoes, aubergines, peppers and winter lettuce, a "natural" patch to remind her of the natural vegetation of the area, then students may be set the task of laying out the garden, given the slope of the land, the location of the house, the position of the well, the nature of the soil, the direction of the *Mistral*. Again, there will be no correct answer, only a variety of answers, some better than others from a practical point of view, some more aesthetically pleasing. Students might want to suggest more exotic plants like mimosa and palm trees!

Variations on such themes may be very numerous. Readers are referred to the pages of *Teaching Geography* and to such exercises as "Motorway Construction" by Eleanor Rawling in *Teaching Geography Occasional Papers, No. 26.* A large number of problem situations may be derived from planning new towns or rehabilitating old towns, or planning new farms and villages as in the Netherlands polders. For

example, a new town with few recreational facilities may decide to set up a sports centre. The problem is one of deciding on its location, given various constraints, such as the position of various residential areas and the nature of the transport network. The improvement of a decaying mining and industrial centre may be the task set to a group of students, as in the Murkyville exercise in *Classroom Geographer* (May 1974).

Sometimes the problem set is that of a mismatch between what is found and what is expected by the students and the explanation of this may lead students to re-examine certain preconceptions or even prejudices. This was a technique used by the American High School Geography Project in shaking up American students' conceptions that black Americans tended to live in poor residential areas. Evidence was provided of a poor and a well-to-do residential area in New Orleans and students were asked to suggest who lived in which. The stereotyped answers proved wrong on further investigation. This led to discussion as to why this should be.

An aspect of problem-solving strategies is the use of simulation and games which have been referred to already at the beginning of this chapter. Much has already been written on this aspect of teaching strategies and various books of geographical games and simulations have been published,[33] consequently little will be written here. Suffice it to say that most teachers and students who have used such strategies agree on their motivational effect. Pupils and students are keen to get involved in such games and simulations. Their ultimate educational effect is less certain, but this is true of almost any teaching strategy. It is so difficult to isolate the variables involved in any teaching situation, that to state that a given outcome is directly the result of a particular teaching strategy is almost impossible to affirm. The attitudes of teachers and students are probably as good a guide as any to the success of any teaching strategy.

The distinctions made between simulations and games are largely matters of degree. All relate to simplified artificial conditions. At one end of the spectrum is the simple role play in which students are given roles to play in a situation in which a decision needs to be taken, say about the cutting of a new road link in a town. The purpose here is to get each student to argue out his case from the point of view of the person whose role he is playing, be this that of the borough engineer or borough treasurer, and for the group to come to a decision. Such role play enables students to practise reasoning and persuasion but also the art of compromise. It may also show them that a forceful personality on a panel may get his own way even if his proposals may not be the most reasonable. A full-scale simulation exercise may involve a more elaborate process of simulating all the stages in a planning proposal for, for example, the sports centre for the new town mentioned earlier. A game, on the other hand, is played with definite rules and an outcome which implies that some individuals or groups win or benefit to a greater extent than others. The Iron and Steel game and the Oil Exploration game are of this type. In some cases, chance factors are introduced to simulate the stochastic element which may occur in many situations. The example which follows was devised by Mrs

Teresa Tunnadine of Minchenden School.

The location of an oil refinery: a simulation 5th Form (16 years old)

Objectives

1. To recognize and evaluate the factors needed in making general locational decisions, industrial location decisions and those specifically related to the location of an oil refinery.
2. To attempt to synthesize and revise information earlier covered by the class in an oil project.
3. To encourage group activity and group decisions and the co-ordination of ideas.
4. To develop written skills. Well laid out and logically structured reports are a necessary part of this simulation.
5. To develop the ability to "argue" soundly orally and on paper.
6. To develop the skill of selecting and ranking the most important facts from an information source.
7. To develop environmental awareness.
8. To introduce the problems of conflicts in land use.

Procedure

Introduction of topic:

  (a) Which factors are necessary for locating an oil refinery?

  (b) Class discussion to clarify the responsibility of each member of the board.

Divide class into groups of four (three, without a chairman if odd numbers).

Hand out Information/Instruction sheets.

Five minutes for class to read over instructions and ask any questions.

20 minutes for each to write their individual reports—without conferring.

Board meeting.

Time to compile company plan.

Each group in turn (chairman) to present their plans.

Voting individually on which site is most suitable.

Chance cards (see below).

Set homework/conclude.

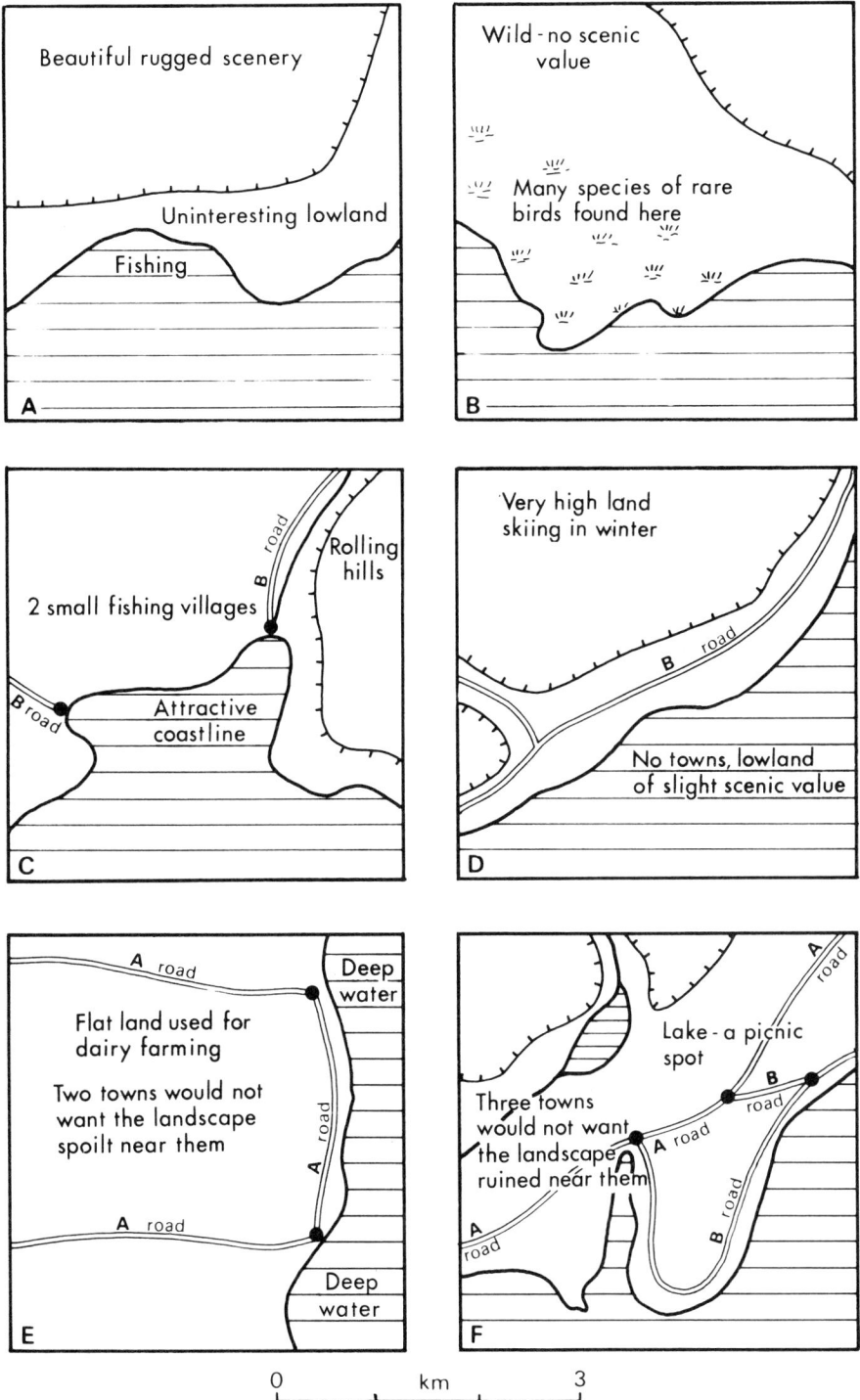

FIGURE 4.28. Possible sites for an oil refinery: environmental planner's maps

Information sheets

Chairman

You are the chairman of an oil company. The Government has announced that it is offering a large grant for building an oil refinery on the east coast of Scotland. Your company has decided to produce a plan to offer to the Government.

The Board is made up of four members. The other three are the economist, the constructional engineer and the environmental planner. Each will be proposing the best site for the oil refinery from their own specialized viewpoints.

It is your job to:

(a) Look at the sites from a general viewpoint; take account of the environmental, economic and constructional factors.

(b) Make notes on the site you consider to be the best and explain in them why this is so.

(c) Listen to the proposals made by the other three members and with them come to a joint conclusion about where the oil refinery would be best located, taking into consideration the environmental, economic and constructional factors.

On behalf of the company you must then put forward the proposal to the Government. It must be supported with reasons for choosing one site and not the other five.

When all the companies have submitted their proposals to the Government it will consider them and decide which it considers to be the best.

There will be time for a general meeting before a final decision is made.

Environmental Planner

You are the Environmental Planner of an oil company. You are one of the board of Directors.

The Government has announced that it is offering a large grant for building an oil refinery on the east coast of Scotland. Your company has decided to produce a plan to offer to the Government with the hope of being the company to have their plan accepted.

The board is made up of four members. The other three are the chairman, the economist and the constructional engineer.

Your job is to look at the six possible sites and advise the board of directors which site you consider to be the best.

You must prepare a list of the reasons you had for choosing one site and not choosing the others.

The company must eventually prepare a joint report on the site it collectively decided upon. This must be put before the Government. Only one plan will be accepted. The plan must therefore be as comprehensive and logical as possible and show it has considered the matter from all viewpoints; i.e. environmental, economic and constructional.

You should consider:

(a) Attractive scenery should be maintained at all costs.

(b) Facilities for skiing, hiking, fishing etc. should not be reduced.

(c) The refinery should be built as far away from towns and villages as possible—especially tourist towns. It might pollute the air and look unattractive.

(d) You should aim to conserve wild animals and plants at all costs.

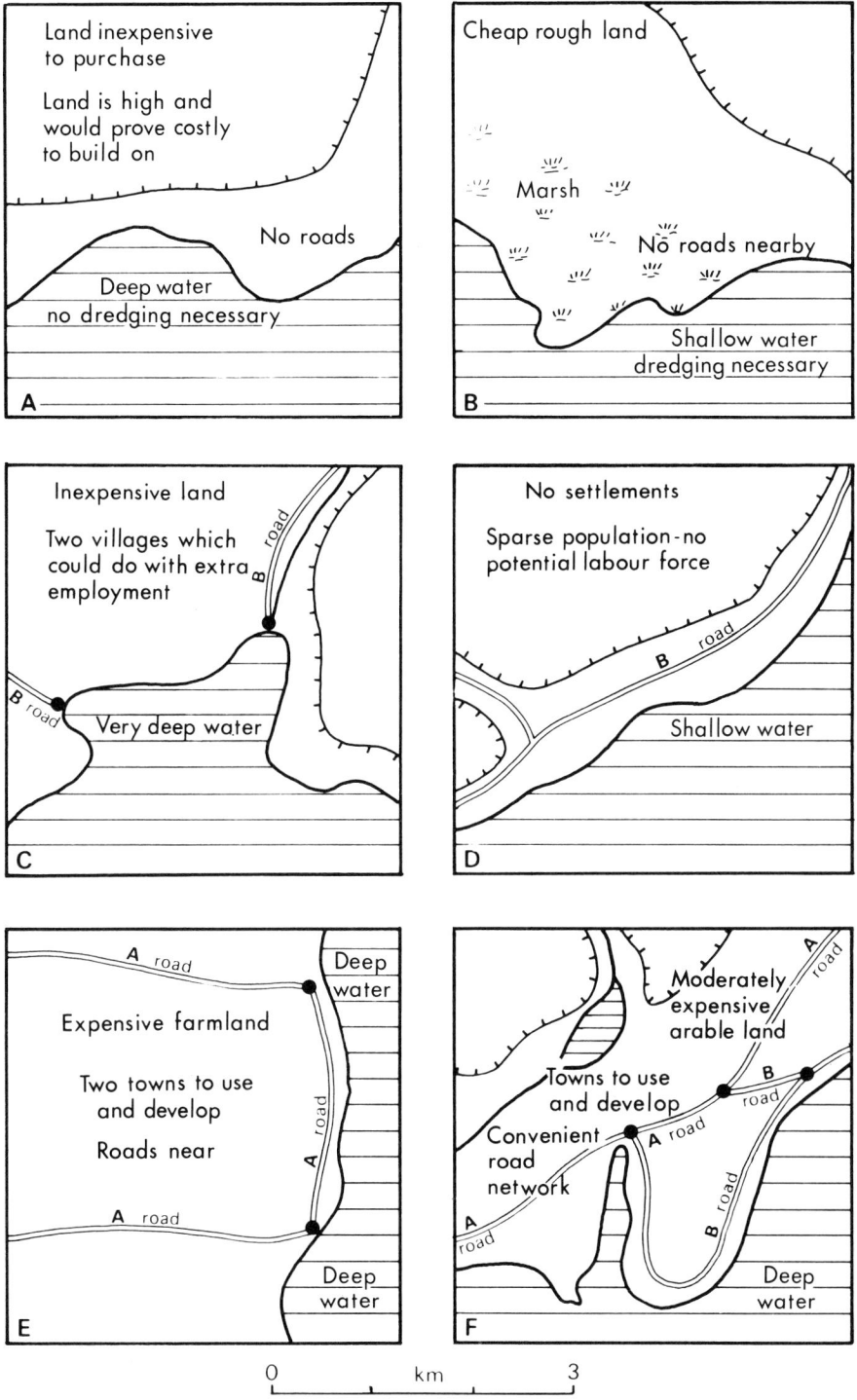

A
Land inexpensive to purchase

Land is high and would prove costly to build on

No roads

Deep water no dredging necessary

B
Cheap rough land

Marsh

No roads nearby

Shallow water dredging necessary

C
Inexpensive land

Two villages which could do with extra employment

B road

B road

Very deep water

D
No settlements

Sparse population-no potential labour force

B road

Shallow water

E
A road

Deep water

Expensive farmland

Two towns to use and develop

Roads near

A road

A road

Deep water

F
A road

Moderately expensive arable land

Towns to use and develop

B road

Convenient road network

A road

A road

B road

Deep water

0   km   3

FIGURE 4.29. Possible sites for an oil refinery: economist's maps

Economist

You are an economist employed by a leading oil company.

The Government has announced that it is offering a large grant for building an oil refinery on the east coast of Scotland. Your company has decided to produce a plan to offer to the Government.

The board is made up of four members. The other three are the chairman, the environmental planner and the constructional engineer. Each will be proposing the best site for an oil refinery from his own specialist viewpoint.

Your job is to look at the six possible sites and advise the board of directors which site you prefer and why.

You must prepare a list of reasons you have for choosing one site above the others.

The company will eventually prepare a joint report on the site it has chosen collectively. This must be put before the Government. Only one plan will be accepted. The plan must, therefore, be as comprehensive and logical as possible and show it has considered the matter from all viewpoints; i.e. environmental, economic and constructional.

You should consider:

(a) The relative cost of the land; how much it will cost the company to buy or hire the land.

(b) Look at the present state of the land. Is it suitable for building on or would a lot of work, involving time and money, be needed? e.g. draining marshy land or levelling hilly ground.

(c) Whether it would be near enough to existing towns and villages to provide work for the people living in the area, which is desperately needed in this part of Scotland.

(d) Is the site well located near road or rail links?

Constructional Engineer

You are the Constructional Engineer employed by a leading oil company as one of its directors.

The Government has announced that it is offering a large grant for building an oil refinery on the east coast of Scotland. Your company has decided to produce a plan to offer the Government with the hope of being the company to have their plan accepted.

The board is made up of four members. The other three are the chairman, the economist and the environmental planner.

Your job is to look at the six possible sites and advise the board of directors which you consider to be the best.

You must prepare a list of the reasons you had for choosing one site and not choosing the others.

The company must eventually prepare a joint report on the site it collectively decided upon. This must be put before the Government. Only one plan will be accepted. The plan must therefore be as comprehensive and logical as possible, and show it has considered the matter from all viewpoints; i.e. environmental, economic and constructional.

You should consider:

(a) Where there is enough firm, flat land to build a refinery on.

(b) Is there enough land to expand the factory should this prove necessary at a later stage?

(c) Is the coastline suitable and the depth of water sufficient to receive large tankers especially as they may become larger over time.

(d) Would roads need to be built to link your site to the existing network?

(e) If roads were built this would be costly and time consuming.

(f) The location of the refinery should be a sheltered one, so the oil tankers can come and go freely without gales interrupting free passage of the import and export of the oil.

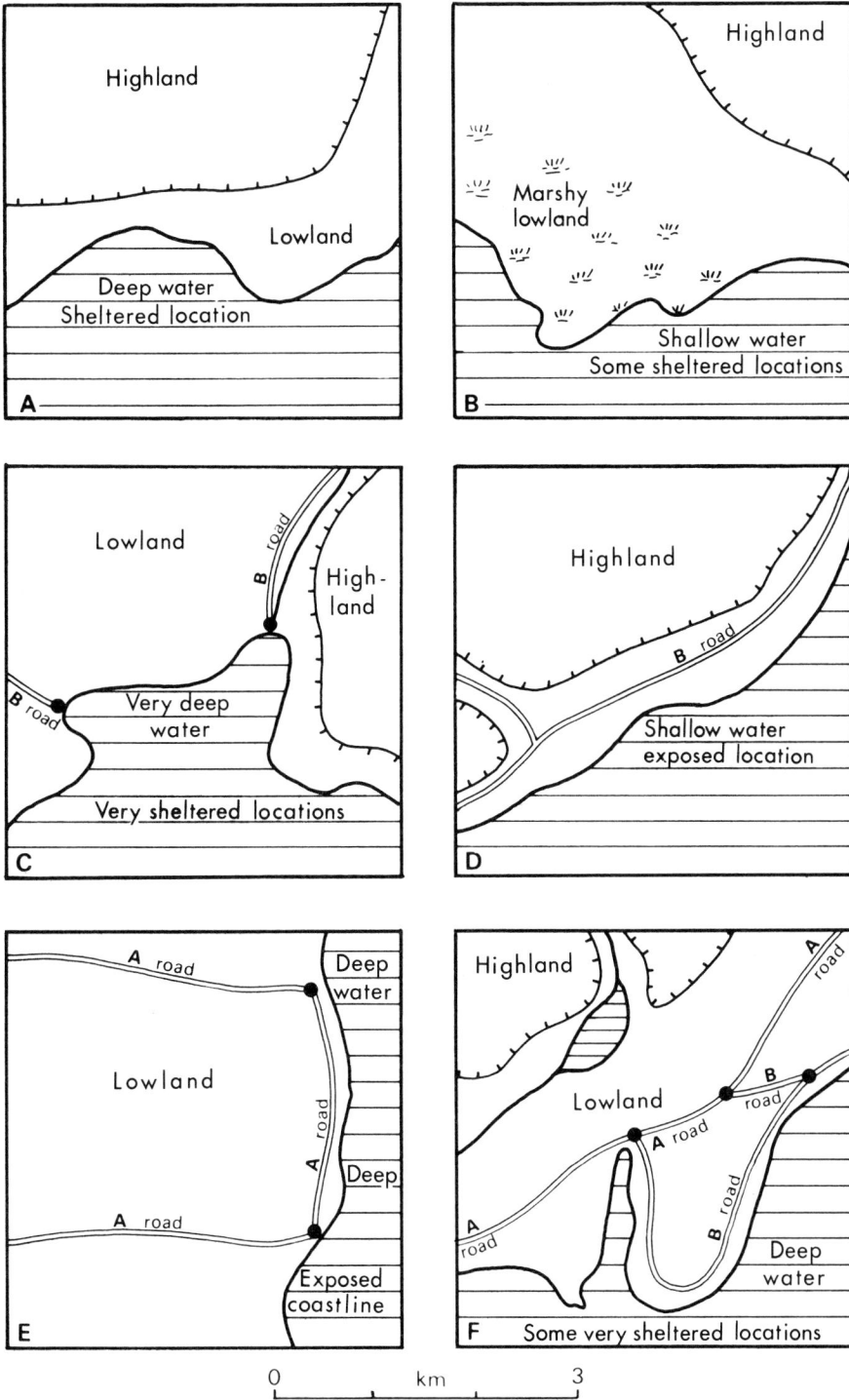

FIGURE 4.30. Possible sites for an oil refinery: constructional engineer's maps

Homework:
- (a) List in order of importance the locational factors you consider to be necessary considerations when locating an oil refinery.
- (b) What are the chance factors (in the real world) that could stop your plan being accepted?

Chance cards:

To round the game off if similar plans are put forward:

1. You have been offered a better contract with Saudi Arabia. Withdraw your plan immediately. There is more profit to be had elsewhere.
2. There has been a change of top management. You are directed to withdraw the plan as company policy is due to change.
3. Your stocks and shares have done very badly. You have not the money to put towards building the refinery. Withdraw your tender immediately.
4. Your refinery in Holland has blown up. All company money is needed to rebuild. Withdraw plan immediately.
5. You are due to merge with company two. No more business can be carried out until this has been carried out. Advice is to withdraw the plan.
6. Your company has been nationalized. As such, the Government has opted to give you the contract to build the refinery.

Clearly simulation and games should be used occasionally rather than frequently otherwise their novelty will soon pall. It is also important to bear in mind that the teacher should spend a little time making sure the educational objective of the exercise has been grasped by the students. The follow-up of such a teaching strategy is not facilitated by the tendency in many schools to give geography one double period (1 hr 20 min) per week, especially in the lower school.

5 Evaluation in Geographical Education

When in the early 1970s the evaluation of geographical education was discussed, it tended to be assumed that what was at issue was the extent of what pupils and students had learned and the means of finding this out. More recently the term evaluation has tended to be associated not just with student learning, but with the whole curriculum. Thus, the term curriculum evaluation refers to the range of processes involved in assessing the worth of the instruction given. It is concerned as much with assessing the process of instruction as with the knowledge, skills and attitudes that pupils and students have acquired. Consequently, it is proposed to examine first the evaluation of instruction and second the evaluation of student learning.

5.1 THE EVALUATION OF INSTRUCTION

It is necessary to go briefly into the issues which have exercised evaluators since the early 1960s. The education process has been under study by philosophers, psychologists and sociologists, as well as by those concerned with the curriculum subjects, but in the field of evaluation it was the psychologists who had contributed the greater amount of research and developed a considerable number of techniques. British and American psychologists were, in the majority, trained in a way which placed value on the experimental and quantitative methods of the natural sciences. It is not surprising, therefore, to find that in considering the educational problem of evaluation, they should have developed a model of the educational process which lent itself to measurement. Broadly, this model, which may be termed the input/output model, visualized the educational process as one in which teachers (and resources) with certain objectives in mind transformed the

pupils and students from a state in which they were ignorant of certain knowledge, skills and attitudes to a state in which they had acquired these. Thus, what was important was to measure the gains in knowledge made by the student. To this end, psychologists who devised evaluation instruments saw the process as one of giving students pre-tests to measure their knowledge prior to instruction, to proceed with a series of teaching units and then to measure by a "post-test" the new state of knowledge of the students. The difference in scores between the pre-test and post-test would give a measure of the gains in knowledge attributable to the instruction. The efficiency of instruction might also be gauged, since techniques of teaching which yielded the greatest gains would presumably be judged to be the most efficient.

Whilst at first sight, this seems an eminently rational approach and appeals to common sense, it has certain flaws which need to be examined. It is necessary to state, however, that in putting into question the input/output model of the education process, I am not arguing that all the instruments of evaluation evolved by educational psychologists are no longer of any use. The baby must not be thrown away with the bathwater!

Let me outline what seem to be some of the problems of the input/output model. First, it relies heavily on the ability of teachers to define very precisely the educational objectives of each teaching unit devised. Yet we have seen in Chapter 3 that whilst some objectives may be reasonably clearly stated, others may not be so easily pre-specified and that in any case there are many unintended outcomes to the education process. Relying too closely on the measurement of performance on objectives could easily lead to a concentration on low-level objectives because these can be easily defined. Thus, it is easy to measure whether a student knows or does not know that Asunción is the capital of Paraguay, but much less easy to measure whether he fully understands the value and limitations of the Weber model of the process of determining the most economic location of an industrial enterprise. Secondly, the model assumes that objectives are given, or at least stable, over a long period of time. In reality, educational objectives in geography, as in other subjects, tend to be in a state of flux; indeed the very concept of curriculum development assumes that those concerned are constantly seeking new objectives and new educational processes. Thirdly, the idea that the efficiency of an educational process can be measured by gains in the knowledge of students assumes that, as in laboratory experiments, all other variables remain constant. This, of course, if it is not the antithesis of reality, is far from the conditions which prevail in the average school, where teachers themselves may vary in style and approach, where environmental conditions may change and not least, where the social psychology of a class will vary from day to day if not from hour to hour.[34]

The upshot of all these doubts is that evaluators are now much less certain that curriculum evaluation may be undertaken solely by reference to an input/output model. They tend to see curriculum evaluation as less dependent on scientific measurement and relying more on judgements made on the basis of observations of the whole educational process, including the planning process. Thus, curriculum evaluation tends now to be concerned with first a broad evaluation of the process of instruction and secondly the evaluation of student learning.

How may the process of instruction be evaluated? This may be done either by professional evaluators employed by curriculum development projects whose function is to monitor the way the curriculum is being developed. Thus, the History, Geography and Social Science 8–13 Project employed Keith Cooper as its official evaluator, and he fed back to the team his views of what was in fact happening in this particular case.[35] Or in the normal process of teaching in schools, teachers from time to time can step back from the hurly-burly of teaching and attempt to evaluate what has been happening to their curricula. In both cases, though it is easier in the case of the professional evaluators, some means of observing the way in which the

planned curriculum is in fact put into practice must be found. This involves teachers in being tolerant of being observed in action and being willing in open discussion to evaluate a process in which they have been intimately involved. Clearly, this means a change of stance on the part of teachers away from an authoritarian one to a much more open-ended one, in which both objectives and teaching strategies are seen as open to adjustment in the light of evaluative discussions. The more open classroom situations of geography departments in comprehensive schools where teachers may be grouped in classrooms around a resources area, may lead to much more inter-classroom observation and mutual discussion than was the case in the past. How far the observation of teacher behaviour needs to be formalized may depend on the predilection or hunch of the evaluator as to what problem he sees as needing particular attention, but observation schedules may be devised such as that originally conceived by Flanders.[36]

The classroom process is also one which is monitored by the pupils and students, and it is possible to administer questionnaires for those in the upper classes of secondary schools in which students are asked what they thought of particular teaching units. For example, in the case of the activities involved in the teaching units on hydro-electric power it would be possible to set out a questionnaire as shown on the next page.

Teachers may also fill in questionnaires on teaching units so that a department may find out how useful its members found particular units devised for the whole department. This allows for modifications to the units to be made in future years. Similarly, if a department runs its curriculum on the basis of a course in geomorphology, followed by a course on settlement, followed by a course on the economic geography of manufacturing and so on, students may also be asked to indicate how they benefited from each course.

This process of evaluating how the geography curriculum has fared in practice has been given the name "illuminative evaluation" since it attempts to shed light on the whole curriculum process rather than attempt precise measurements of knowledge gained by students.

Evaluation of teaching units on HEP

Please tick the box which most represents your feelings on the unit.

	Strongly Agree	Agree	Disagree	Strongly Disagree
1. I enjoyed working in groups.	☐	☐	☐	☐
2. I learnt a lot from the individual cases.	☐	☐	☐	☐
3. I found the summary table useless	☐	☐	☐	☐
4. The application exercise on the R. Fiume dam site was too easy.	☐	☐	☐	☐

5.2 THE EVALUATION OF STUDENT LEARNING

Part of the evaluation process is still the evaluation of the knowledge, skills and attitudes that students have acquired during the process of education. In British education and in many other European countries, the process of student evaluation is formalized with numerous tests and examinations. These have undergone changes in the 1960s and 1970s and the methods employed are now more varied and more sophisticated than they were. It is realized that the method of evaluation must closely parallel the behaviour it is intended to measure. For example, at one time the essay-type question was almost the only one used in examinations in geography, although its function was mainly to enable students to recall knowledge and organize it in a way suitable to the requirements of the question. Today the range of questions used is much wider, just as there are many modes of examining.

Let me first be clear about the terminology used. A *public or external* examination is one organized by such boards as the General Certificate of Education or Certificate of Secondary Education Boards which are officially approved by the Department of Education and Science. Teachers will be aware that the DES had accepted that a common system of examinations at the 16 + level be instituted in order to avoid the dichotomy between CSE and GCE courses, which complicated school timetabling and promoted insidious distinctions between students. An *internal* examination is one run by a particular school or institution for its own students. Many schools have once or twice-yearly examinations to monitor pupils' and students'

progress. The process of examining may be divided into two main categories: (1) course work assessment, in which work done by students during a course is formally assessed and contributes to the final grade given; (2) a formal written examination or test during which a student is expected to answer a set number of questions in a limited time period. Most public examinations are still of the latter type, though a number of CSE and GCE examinations in geography, such as that for the "Geography 14–18 Project" run by the Cambridge Local Examinations Syndicate, combine both course work assessment and a formal written test. Within the formal written examination, the kind of questions set may vary considerably, from the straight essay questions, through the structured data response essay questions, to the objective-type questions. Each has a function which makes it suitable for testing certain types of behaviour and unsuitable for others. For example, the type of question which states:

The area of town which consists mainly of shops, offices and public buildings is known as the:

Place tick in the correct box

(a) industrial quarter	☐
(b) residential area	☐
(c) central business district	☐
(d) public administration area	☐
(e) twilight zone	☐

is a question which seeks to find out whether the

student has a knowledge of the appropriate terminology used to describe the commercial heart of towns or that he is sufficiently discriminating to reject what are inappropriate terms and thereby choose the correct one. But it does not enable the student to argue his case out or indicate that (d) describes part but not the whole of the central business district. Consequently, an examiner must decide what it is he wishes to test and what the purpose of the examination is.

In internal examinations, monthly or termly tests may be intended to assist the teacher in diagnosing whether students have understood and are able to use certain concepts, principles or skills which he has taught. In such a case he may give an objective test which measures the understanding of certain concepts and a number of application exercises which will test whether these concepts may be used in new situations. The expectation would be that most students would have mastered the ideas and therefore marks in the test would generally be high. If they were low, this would be an indication that the concepts were too difficult or had not been successfully taught though the students were deemed capable of assimilating them. I would argue that most school tests ought to be of this type and that examinations whose sole purpose is to produce a rank order in a subject are not particularly valuable. With the results of a diagnostic-type test, the teacher may take remedial action where necessary, which gives the test a positive function.

Public or external examinations have the task of guaranteeing to the public at large and to employers in particular, that a candidate who has passed the examination with a certain grade has reached an agreed standard of performance in the subject concerned. Given this objective, the grading of students assumes more importance than in internal examinations. Nevertheless this does not absolve the examiners from thinking out carefully the kind of performances by students that they are guaranteeing. Thus, in geography it would appear that formal examinations are monitoring such abilities as:

1. the ability to recall geographical terminology and some relevant facts;
2. the ability to understand and apply such terminology in new situations;
3. the ability to analyse a problem and suggest a

solution based on the use of previously learned concepts, principles and techniques;
4. the ability to use map information and infer spatial relationships from that information;
5. the ability to plot certain distributions on maps;
6. the ability to use various kinds of graphic information;
7. the ability to draw graphs and diagrams illustrating quantitative relationships.

The techniques used to evaluate such performances will vary according to the nature of the performance. The evaluation of (1), (2) and (4) is probably best done by objective tests covering a wide area of the syllabus; the evaluation of (5), (6) and (7) is best done by data-response questions, whilst the evaluation of (3) is probably best undertaken by an essay-type question. Much depends on the intellectual level of the knowledge or skill being evaluated, whether an essay or data-response question is best in any given situation. The more open-ended a situation, the more an essay-type question is appropriate. On the other hand, the marking of such questions is notoriously difficult to standardize, and to ensure parity between examiners is an almost impossible task.

My own view is that at the 16-year-old level, most questions should be of the objective and data-response type. An objective question is, of course, only objective in its marking, since there is only one correct response, but decisions about the kind of knowledge tested is subjective. A danger with objective-type questions is that since it is much easier to set factual or terminology-recall-type questions, these may tend to dominate in a given examination. It is the responsibility of the chief examiner to ensure that most questions involve some thinking on the part of the candidates.

Here are a couple of examples; the first of a data-response question with some latitude in the way the candidate answers the question, and the second of an objective-type question where there is little latitude given to the candidate but he must think out what the correct answers are:

1. Examine photographs (Plates 15–19) and the land-use map of the Netherlands (Fig. 5.2).
   (a) Indicate in which land-use zone each photograph was taken.
   (b) Explain what physical factors of relief and

PLATE 14 Aerial photograph of a peninsula on the south-east coast of Greece near Athens

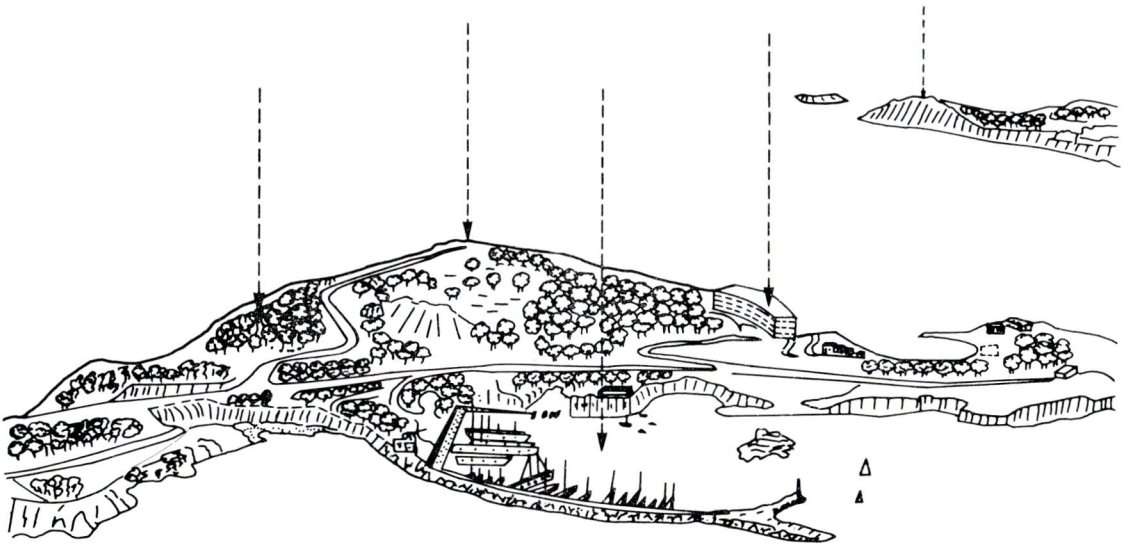

FIGURE 5.1. Sketch of the peninsula shown in Plate 14

Photograph by Bart Hofmeester, Rotterdam

PLATE 15 Landscape of the Westland (Province of South Holland)

Photograph by Bart Hofmeester, Rotterdam

PLATE 16 Landscape of Broek op Waterland (Province of North Holland)

PLATE 17   Reclaimed land in the Netherlands   *Photograph by Bart Hofmeester, Rotterdam*

PLATE 18   Landscape near Almen (Province of Gelderland)   *Photograph by Aerophoto Teuge*

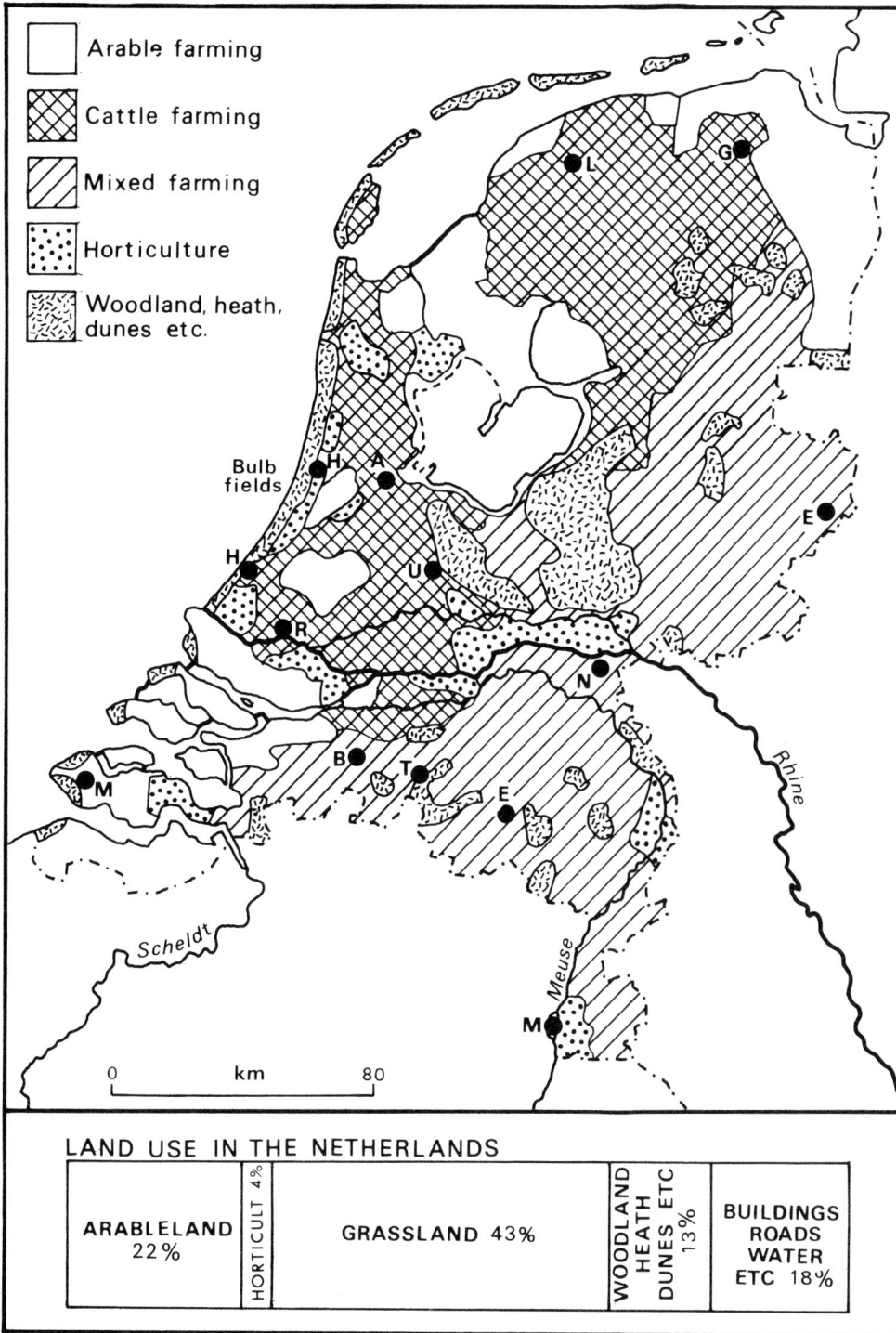

Legend:
- Arable farming
- Cattle farming
- Mixed farming
- Horticulture
- Woodland, heath, dunes etc.

Bulb fields

Rhine

Scheldt

Meuse

0   km   80

LAND USE IN THE NETHERLANDS

ARABLE LAND 22%	HORTICULT 4%	GRASSLAND 43%	WOODLAND HEATH DUNES ETC 13%	BUILDINGS ROADS WATER ETC 18%

FIGURE 5.2. Land use in the Netherlands

Photograph by Wim Steffen, De Steeg

PLATE 19   Heathland at the Veluwe (Province of Gelderland)

soil help to explain differences in land use in the Netherlands. Draw a section from E. to W. across the Netherlands to illustrate your answer.

(c) Describe what major actions human beings have taken to modify land use in the Netherlands in historic times. Why do you think the effort and investment required were undertaken?

2. Study the physical map of the Netherlands (Fig. 5.3). Place a tick in the box opposite the correct answer:

(a) If the sea were to breach the dykes in the Netherlands and sea level were to rise by 1 metre which of the following towns would still be above sea level:

    (i) Amsterdam   ☐
    (ii) Rotterdam   ☐
    (iii) Leewwarden   ☐
    (iv) Nijmegen   ☐
    (v) Middleburg   ☐

(b) The sandy soils north of the R. Rhine are split into an eastern and western area, by an area of non-sandy soils, this is because:

(i) Men have bulldozed out the sands between the two areas to make a passage for the Ijssel river.   ☐

(ii) The ice sheet which deposited the sand did not deposit sand in the Ijssel valley.   ☐

(iii) The Ijssel river has worn away the sand and deposited alluvium.   ☐

(iv) The two sandy areas were deposited by two separate ice sheets which came from the north but never met, leaving an unglaciated area between them.   ☐

(v) Farmers have so improved the soils with manure and fertilizers in the Ijssel valley that little of the original sand is left.   ☐

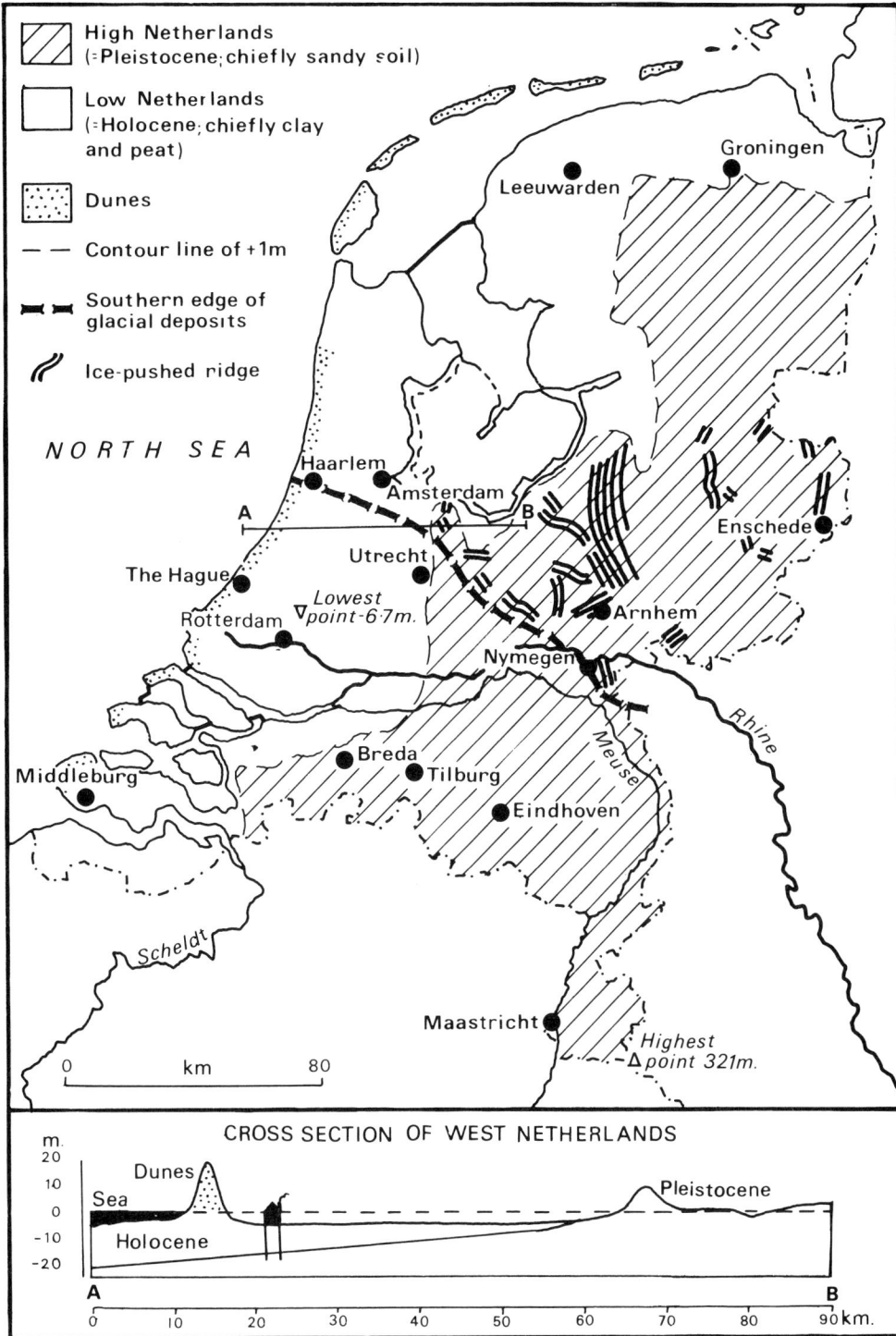

FIGURE 5.3. Physical map of the Netherlands

(c) There is an area of sandy soil in the southern Netherlands, yet this is south of the southern edge of glacial (ice sheet) deposits, this is the result of:
  (i) Desert conditions in the past. ☐
  (ii) Man's transport of sand from the north to the southern Netherlands to improve the soil. ☐
  (iii) Westerly winds blowing sand in from the coastal areas. ☐
  (iv) The area used to be covered by dunes which have been flattened out. ☐
  (v) Meltwater from the ice sheets spreading sands and gravel south of the edge of the ice sheet. ☐

Objective-type questions need to be tried out with groups of students to find out whether they work or not. Those involved in drafting them for official examinations find it worthwhile to calculate the facility and discrimination indices for each item. Teachers will probably be able to judge from the response of their pupils whether an item is suitable or not. Devising such items can be time consuming; consequently teachers would be well advised to store a bank of such items to be used from time to time with different students. Care should be taken when using a test item again that the data in it are still valid, as questions may be based on statistics which are no longer applicable, as for example Britain's sources of energy which are liable to change rapidly over a short period of time.

5.3 CONCLUSION

Evaluation is part of the curriculum process, but it should not dominate this process. There is inevitably a danger, within an education system that involves official examinations, for the whole process of education to become subservient to that of evaluating student learning. The teacher should always refer back to the aims which he has set himself in teaching geography and ascertain whether the kind of evaluation he is doing contributes to these overall aims or not. If it does not, he should have no hesitation in abandoning that kind of assessment.

6 Resources For Teaching Geography

This section will be relatively short since much of the information which teachers require on resources may best be placed in an appendix. If we return to the curriculum process diagram (Fig. 1.1) we can see that resources may be required for objectives, content, teaching strategies or evaluation. I will, therefore, take each in turn, and outline the kind of resources available.

6.1 RESOURCES ON AIMS AND OBJECTIVES

Under this heading, the teacher may require to read books and articles which may help to clarify or modify his ideas as to what he is trying to do when teaching geography. Clearly, no one engaged in the demanding task of teaching adolescents is going to be continually questioning his aims, otherwise very little teaching may take place, or the direction of the instruction may not be very clear. But it is not an unhealthy situation to reconsider from time to time what one is doing, and to do this satisfactorily it is useful to read recent texts and articles on geographical education, where geography is discussed in the total educational context, and to discuss these with colleagues in the schools and with other geography teachers. Thus, books by Bailey,[37] Boden,[38] Graves,[39] Hall[40] and Marsden[41] published in the United Kingdom discuss in some detail the aims and objectives of geography in education. It is salutory to examine what is being written by those in a slightly different, culture; hence American, German, Australian or Scandinavian texts often present refreshingly different points of view.

To limit the range, let me quote Manson and Ridd,[42] who have put together a series of contributions for the National Council for Geographic Education, which is the American equivalent of the Geographical Association. In reading American texts, it must always be borne in mind that the curriculum context is very different from that in the UK. Few American secondary schools teach geography as a separate subject. Geography tends to be encapsulated in a social studies context, with the result that most teachers who may teach geography courses within that context, will not have been trained in geography. Other sources of information in English on aims and objectives will come from occasional articles in periodicals such as *Teaching Geography, Geography,*[43] *The Journal of Geography,*[44] *Geographical Education*[45] and *Classroom Geographer.*[46] The teacher who wishes to go more deeply into questions of aims and objectives will need to read some of the educational literature on this subject, a lead to which may be given here in the books by Barrow,[47] Hirst,[48] Peters[49] and White,[50] as well as in such periodicals as the *Journal of Curriculum Studies*[51] and the *British Journal of Educational Studies.*[52]

6.2 RESOURCES ON CONTENT

Owing to the vast range of possible objectives which can be arrived at within the orbit of the general aims of geographical education, many teachers welcome some guidance as to what content to choose. By content, I mean the kinds of ideas or concepts, principles or theories which will help to achieve the aims and objectives postulated. To this end Her Majesty's Inspectors for Geography put together a booklet[53] which gives suggestions as to the kind of generalizations which could form the content of a geography course. Similarly, Graves in *Curriculum Planning in Geography*[54] has attempted to outline a source of content arranged in such a way that this content is divided into three stages (Stage I for 11–14-year-olds, Stage II for 14–16-year-olds, Stage III for 16–19-year-olds) and the content for each stage is sub-divided according to certain themes in geography (geomorphic systems, agricultural systems, settlement systems etc.). This scheme assumes that geography may be taught as part of combined studies in the lower secondary school (Stage I) but as a separate subject thereafter. It is based on an ecosystem view of geography as shown on Figure 2.1. Another view of the structure of geography would yield a somewhat differently arranged source of content. A source book for content in human geography from which teachers may derive some inspiration is Haggett's *Geography: A Modern Synthesis*.[55] This by no means exhausts the kind of resources available, since publishers are each year publishing new books on geography which are either texts which bring together the main findings in a field, or collections of articles on a problem or topic. Consequently, teachers are advised to attend the Annual Conference of the Geographical Association which has a large publishers' exhibition of all the major books in geography both at the school and higher education levels.

A simple way of choosing content is to rely on a text book series which may contain a course to 'O' or CSE levels. Whilst this may be a simple solution to the problem, it is seldom completely satisfactory since what the school may need will not necessarily be provided by the text, neither will it be in the order required. Most teachers may find their needs best catered for by using a variety of text books, which they can issue from time to time according to need, to the students concerned. This provides some organizational difficulties in the control of resources which will vary according to the size of the school.

Yet another source of content comes from the published resources of various curriculum development projects in geography. In some cases, the project will have suggested a fairly tight course structure as in the Geography For The Young School Leaver Project,[56] in others the project may have left many curriculum content decisions to the teacher, though if the project is associated with an examination, as in the case of the Geography 14–18 Project, then some tighter content structure may be imposed by the examination syllabus.[57] For the lower secondary school, the History, Geography, Social Science 8–13 Project may provide useful insight into appropriate content.[58] At the sixth-form level, the current Schools' Council Geography 16–19 Project is proving a most useful source of content. Its systematic consideration of objectives; its man–environment framework; its 4 main themes,

1. natural environments: the challenge for man
2. use and misuse of natural resources
3. man–environment issues of global concern
4. organization and management of environments for man;

its A-Level syllabus for the University of London School Examinations Council; and the many teaching units in preparation, all promise a rich source of information for the teacher looking for help in structuring a sixth-form course.

6.3 RESOURCES ON TEACHING STRATEGIES AND INFORMATION

Once the content has been decided upon or indeed in the process of deciding on content, the problem of strategies and information to use will arise. Sources of teaching strategies may be obtained from the books on geographical education cited earlier in this Chapter or in Chapter 4. However, teaching strategies may often be suggested by the kind of information that becomes available to the teacher. For example, the existence of information on Le Havre (as in Chapter 4) may lead the teacher to use an inductive strategy based on a case study to teach about port location and development and port industries. Thus information resources, teaching aids and teaching strategies are often closely linked. I shall, therefore, examine sources of information under 3 headings: (1) documentary resources, (2) audio-visual resources, (3) computer-assisted learning.

6.3.1 Documentary Resources

Apart from text and other books there is a wide variety of printed materials available through various organizations, some of them catering specifically for teachers. It would be tedious to list them all here, and references will be found in the Appendix, but certain examples may be given. First, there are those catering specifically for geography teachers. The magazine *Geo* contains brief informative articles with diagrams, maps and photographs which are aimed at secondary school students and may form the basis of worksheets on various topics. There are also information sheets and booklets which may often be obtained free of charge from the information offices of various embassies and high commissions. For instance, the Japanese Embassy will provide some most useful material on modern Japan including, if required, the organization of day conferences on aspects of Japanese life, economy and culture. Large industrial groups, nationalized or private, such as the British Gas Corporation often have educational materials for schools. Similarly, certain organizations whose purpose is philanthropic may, like the Council for World Development Education, provide useful documentary resources; in the case quoted information and teaching suggestions about development in the Third World. Most of such information has the purpose of presenting the country or industry in a favourable light. However, teachers may use the resources provided in such a way that students examine them in a critical light. For example, the *Daily Express* once put out a pamphlet attempting to demonstrate that (at the time) Britain had a much higher per capita national income than other members of the EEC, and that if we joined the Common Market our per capita income would come down owing to increasing trade with such economically poor partners. By discussing the arguments advanced with students in the light of the theory of international trade (comparative cost advantages) they soon came to the conclusion that if the arguments advanced were correct then no country would ever trade with countries poorer than themselves—an absurd result.

The extensive use of such resources is made possible by the reprographic techniques now widely available in schools. I am referring not only to the long-established spirit and stencil duplicators, but also to photocopying equipment and offset-litho printing. It is possible, by a judicious use of printed extracts, maps, photographs and appropriate instructions and questions, to provide students with attractive resources which form part of a teaching unit on a given theme or problem.

6.3.2 Audio-visual Resources

Teachers will be familiar with the traditional audio-visual aids; the slide and filmstrip projector, the moving film projector and the overhead projector, the last of which is probably the most used form of visual aid in Britain today. The overhead projector is extremely flexible since it can project commercial or teacher-made transparencies, the latter can be made in many colours to reproduce maps, diagrams, printed texts or even photographs. The use of the heat copier means that professionally produced resources may be copied on a transparency for use on the overhead projector. It is possible to use a basic map of a region and by a series of overlays gradually build up a picture of significant relationships, such as distribution of population in relation to relief, to mineral resources and communications. Also widespread is the use of radio programmes, since these may be easily recorded on tape and played back according to requirements. Some are associated with filmstrips to form the Radio-Vision programmes. Most of these programmes are meant to supplement information available in documentary form, by bringing authentic sounds, up-to-date information and discussion by experts on a topic of geographical interest. Somewhat less widespread is the use of television, since to benefit from the programmes produced it is best to record them in colour and reproduce them at a suitable time. Copyright laws limit the number of years for which tapes may be kept. Not all schools have colour video recorders, though the number is increasing. Both the BBC and ITA companies produce television programmes to illustrate themes in geography. An example is the BBC series "Europe From The Air", which combined the advantage of ground-level photography with extensive helicopter filming of important clusters of population such as the Ruhr and the French Riviera. These programmes have the advantage of up to dateness and authenticity as well as being professionally well presented. Information and supplementary documentation may be obtained from the BBC or the ITA companies. Tape-slide programmes are also available commercially though a teacher with appropriate slides may be able to record a commentary and questions useful for his classes. Such home-made programmes may help non-specialist teachers.

Moving films for the 16 mm projector are also extensively used, though these have the disadvantage of rapidly becoming out of date. This is a problem that teachers must be aware of. Familiarity with a good film may lead a teacher to use a film for many years; I am guilty of having done this myself. But imperceptibly the film's visual impact may be lessened by the fact that many of the scenes it portrays were clearly shot many years ago, fashions, cars and buildings may have changed significantly. Further, if it deals with either information or theory, these also may no longer be relevant. Unfortunately, educational films are seldom renewed as frequently as they should be. It is a pity that many of the excellent documentary films shot for television (for example, David Attenborough's "Life On Earth") are not made immediately available for commercial reasons to schools and other educational institutions.

Sixteen-millimetre moving films may be obtained from a variety of sources. First, the local education authority is likely to have a film library and most schools will possess its catalogue. Secondly, a number of national film libraries exist, from which educational films may be borrowed. Thirdly, a number of commercial film libraries exist such as the Encyclopaedia Britannica film libraries. Lastly, various organizations such as the petroleum companies and Unilever have a stock of 16 mm films, many of which are technically excellent and appropriate for use in schools. For example, Unilever has a series of films on the cultivation and production of oil seeds which contain first-hand accounts of conditions in West Africa (for ground nuts and oil palms) or south-west France (for sunflower oil) or the USA mid-west (for soya bean oil). Teachers need to view the films first to decide whether they are worth showing, and if they are to arrange for the way they will be incorporated into teaching units. Unless a film is a technical one concerned with processes (and there are a number of these in the field of geomorphology and climatology), its main impact will be on the emotions. A number of films made in Australia on various aspects of the city illustrate this point well. They use selective rapidly succeeding zoom lens shots of various Australian cities, interspersed with loud or soft music of the "pop" variety and interjections of personal comments of local inhabitants to create a mental image of a decaying inner urban area, of an industrial area, of a genteel suburban district and of a central business district. They may be borrowed from the Australian High Commission. Help on audio-visual aids may be obtained from the National Audio-Visual Aids Centre (address in Appendix).

6.3.3 Computer-assisted Learning

A relatively recent innovation is the use of the computer as a teaching resource in geography. The computer has for many years been used as a means of undertaking rapid calculations in the field of geographical research. Indeed, the so called quantitative revolution of the 1950s in the USA was largely dependent on the availability of computing facilities. Today more and more schools have access to a computer through a computer terminal or schools may have their own mini-computer, a development made possible by the existence of miniature unit printed circuits, the "silicon chip". The terminal (whether linked to a large main frame computer or to a local mini-computer) consists of a keyboard on which may be typed messages giving instructions to the computer and into which data may be fed. By depressing the "return key" on the keyboard the computer returns to the sender the results of its operations. These are either displayed on a video-display unit (VDU) or printed out at considerable speed by a computer print-out machine on paper.

What the computer does depends on the way it has been programmed to operate. In general, its activities in the field of geographical education may be divided into two categories. Either it can be used as a calculator or inter-actively for simulations and games. In the first case it enables a great deal of data to be processed very rapidly. Thus data gathered in the field or in a library may be treated in a standard way to obtain, for example, the standard deviation of a distribution, or the product moment (Pearson's) correlation coefficient can be calculated between two distributions if it is suspected that they might be related, or an analysis of variance may be obtained on data from several groups to see whether any group is apparently very different in its characteristics from the rest. All this is granted that such techniques are appropriate to the problem under investigation. An extension of this use of the computer is the production of

FIG. 5 AFFLUENCE (%CAR)

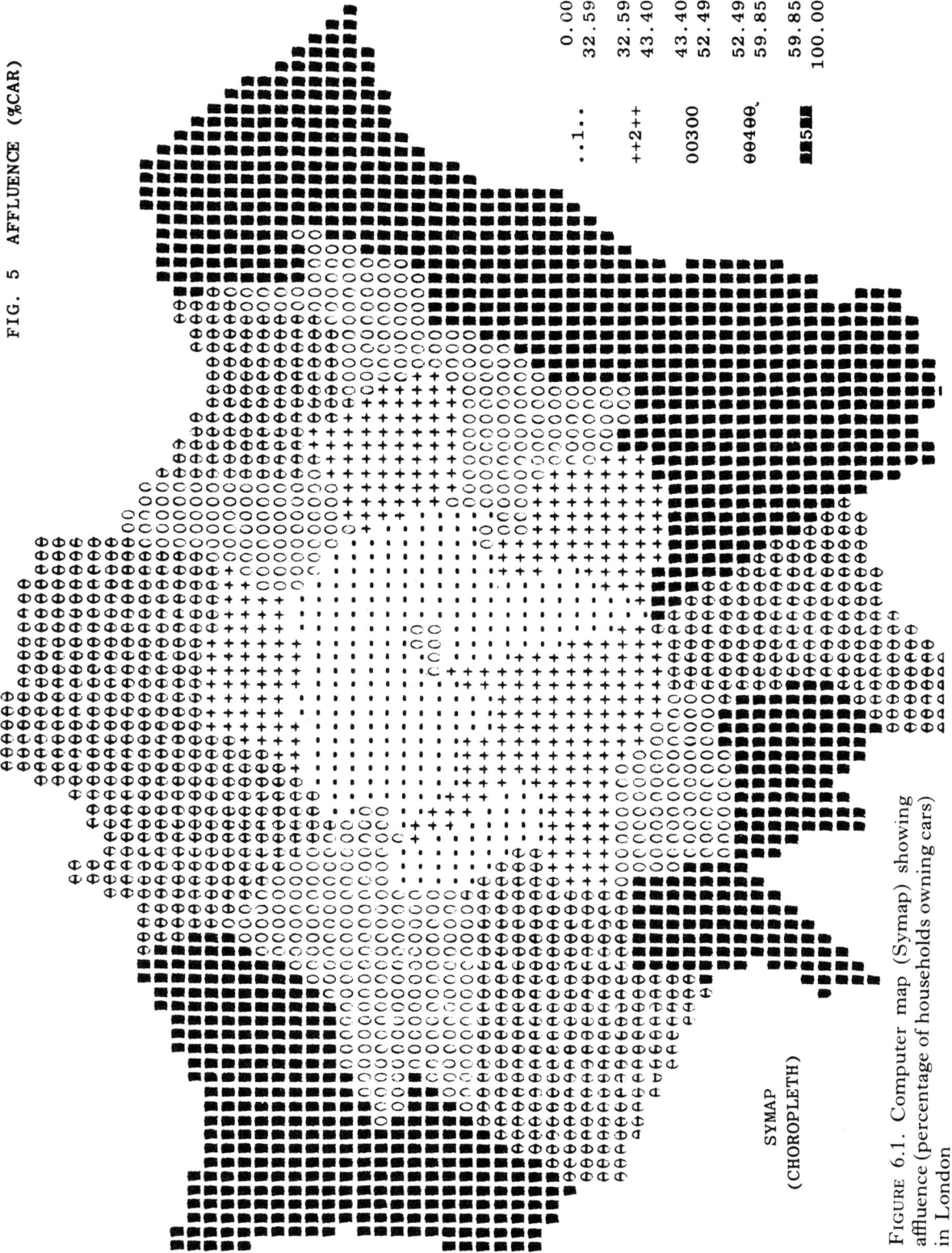

..1..	0.00	32.59
++2++	32.59	43.40
00300	43.40	52.49
00400.	52.49	59.85
▄▄5▄▄	59.85	100.00

SYMAP
(CHOROPLETH)

FIGURE 6.1. Computer map (Symap) showing affluence (percentage of households owning cars) in London

graphics; that is, a distribution may be shown in histogram form, or computer maps of particular distributions may be obtained, for example, the distribution of the percentage of households owning cars (Fig. 6.1). Thus the operation involves calling up the correct computer programme, feeding in the data and obtaining the treated data. In using the computer inter-actively, the computer has been programmed to react in certain ways depending on the instructions given to it. The simplest case to illustrate this is a game designed to teach young children or adolescents the use of grid references. It is called "Hunt The Hurkle", and consists of trying to locate by means of grid references (2, 4 or 6 figures depending on the level of sophistication required), a "hurkle" lost in a large area. When the correct programme has been called up, the student is asked to guess the location. This he taps out on the keyboard; according to the figures typed out, the computer replies with instructions to go north-east or south-west or whatever is appropriate. The student then taps out another reference. This process continues until the student has correctly located the position of the lost "hurkle". Many other simulations have been devised which enable students to use the computer in this way, such as the Chagga Farming Game, in which decisions have to be made as to what crops to plant in an African situation with varying eco-logical conditions (climate, pests, availability of fertilizers etc.). The result is given in terms of yearly crop outputs.

The advantage of the computer in all these situations is the rapidity with which results are obtained; that is, given a set of conditions the consequences are known almost instantaneously, whereas in a situation where laborious calculations would have to be made, the excitement of the situation would be lost.

Information about the use of the computer in geographical education may be obtained from the Geography Department at Loughborough University of Technology, where the Geographical Association's Package Exchange (GAPE) scheme was originally based. The computer programmes which may be fed into the school's own computer may also be obtained from the same source. The Hertfordshire Education Authority and Hatfield Polytechnic have been active in the development of the computer in schools and reference to the authority may put teachers in contact with geography teachers in Hertfordshire who have developed their own programmes.

Although the widespread use of computer-assisted learning is restricted by the availability of computer terminals at present, with the advent of micro-computers it may soon be possible for a greater proportion of students to benefit from this technological innovation.

6.4 RESOURCES FOR EVALUATION

As indicated in the chapter on evaluation, the art of curriculum evaluation is in a state of development and evolution. As a result, resources for the teacher are limited, being more numerous for the evaluation of student learning than for the evaluation of instruction.

Let me, however, begin by indicating those sources which are concerned with the wider view of evaluation. Marsden's book *Evaluating the geography curriculum*[59] is probably the main work in this field, although it deals with other aspects of curriculum besides evaluation. Each of the Schools Council's Geography Curriculum Development Projects publications have sections on the evaluation of instruction, but for a general statement on curriculum evaluation, teachers are advised to consult the Schools Council's *Curriculum evaluation today: trends and implications* edited by Tawney.[60] A problem at present is that many of the texts on curriculum evaluation are theoretical discussions of the issues[61] rather than practical advice as to what to do or what instruments to use. Nevertheless, the art of curriculum evaluation is making progress and in a few years it may be possible to recommend a manual for curriculum evaluation rather than books discussing the problems.

On the evaluation of student learning, resources are much more developed and helpful to the teacher. A general review of evaluation techniques in secondary schools is contained in Deale's Schools Council publication.[62] Most books on geographical education contain sections on examinations and tests, though Marsden's is probably the most thorough in this field. A special chapter on examinations may be found in *New movements in the study and teaching of geography*[63] whilst Salmon and Masterton[64] devote a whole book to objective testing. From an American point of view, Kurfman,[65] the evaluator of the American High School Project, has collected a series of articles on evaluation in geography which was published in 1970. From time to time articles have appeared in *Geography* on objective testing such as those by Hones and Marsden,[66] and publishers have also put out a number of slim booklets of objective tests.[67] These should be scrutinized with the criteria in mind which were outlined in the chapter on evaluation, as many contain too many factual items.

Lastly, current thinking about examinations and tests may be gleaned from the kind of tests and examination papers put out by the various examination boards for the GCE and CSE. The addresses of such boards may be obtained from the back pages of *Geography*[68] which publishes annually details about these boards, their chief examiners and geography subject panels. Further, some boards, such as the University of London School Examinations Council, have research departments which publish from time to time information papers about examination procedures and techniques.

7 Curriculum Planning and Development

I now return to the subject which we considered in Chapter 2, namely the geography curriculum, but this time from the point of view of the curriculum planner and developer.[69] I am assuming that most teachers will want to undertake some detailed planning of what they intend to teach and that, if the analysis of the curriculum problem set out in Chapter 1 is valid, the curriculum is a dynamic system which will change or develop from year to year. Thus, planning the curriculum involves not the formulation of a static body of knowledge to be imparted to students, but devising a system which enables the teacher to devise learning experiences, which will change from time to time, to achieve certain objectives, which again may change over time, but which objectives will contribute to the overall aims of geographical education, which yet again may change, though more slowly than the specific objectives. Changes are brought about by many influences, including the results of evaluation procedures, which will tend to affect learning experiences and objectives, but also by the changing demands of society and the changing values of the education system and the teachers.

The position of geography in the school curriculum as revealed by the HMI's survey in *Aspects of Secondary Education in England*,[70] is one which has been extant for many years. In general geography is taught as part of a common core curriculum for the first three years of the secondary school. It is not clear, however, just what proportion of schools teach geography in those years as part of a wider combined group of subjects, whatever label is given to such a group—Integrated Studies, World Studies, Social Studies, or Humanities. My own guess is that about 25% of secondary schools teach "combined subjects" in the first three years. It is in such situations that team teaching is most frequently met. In those schools a special responsibility lies with the geography department to ensure that the curriculum contains those elements of geography which are essential to the total curriculum and which teachers will want to develop in later years. It may be stated incidentally that the HMI's survey found little evidence that Integrated Studies promoted a holistic view of the curriculum. Most schools then, include geography in the 4th and 5th years of the secondary school course as an option, rather than as part of the core curriculum. The arrangement of options is variable, but approximately two thirds of all schools arrange their options in blocks. This means that the student may choose one subject in each block, for example from a block combining Humanities, Biology, Art and Geography. This does not necessarily mean that if, for example, he chooses biology he cannot do geography, for another block may also contain geography. In some comprehensive schools the students are 'banded'; that is, students are divided into three groups in each year corresponding roughly to three ability levels. In such schools it is not unusual for the option blocks to be different for each band, but in such cases geography is often offered in two or even all three bands. Another form of organization is to set subjects; that is, to ensure that in one band all periods of one subject are timetabled together so that the students may be arranged in a number of sets, again each set corresponding to a certain attainment level in the subject. Thus high achievers may be taught together, low achievers can be given special attention. In practice geography is seldom among the subjects which are taught in sets. Conversely, in schools where it is policy to teach some students in mixed ability groups, then geography is often a subject taught in such groups, particularly in the first three years. Teaching in

86

mixed ability groups is not easy as in effect it is necessary to undertake small group or even individualized teaching. This is probably why when examination courses are planned they are seldom planned with mixed ability teaching groups in mind.

The existence of an option system is beginning to be seen as not as beneficial as it once was. Educators are beginning to question whether it is good that a substantial proportion of students should leave school with little understanding of their natural and social environment. Students in the lower bands are often at a greater disadvantage through the option system than those in the upper bands, since few options tend to be open to them.

7.1 CURRICULUM PLANNING

Few geography departments are in a position to plan their geography curriculum from scratch. They are usually faced with an existing syllabus even if this is only the syllabus of an examination board. Consequently most teachers go through a process of gradually modifying their curricula as opportunity and need arise. Nevertheless, it is useful to have an overall conception of the way the curriculum may be planned, so that modifications can be brought about in a rational manner.

It may be useful at this stage to clear up some of the terminology used in connection with the curriculum process. The term syllabus has already been used and traditionally this refers to a list of content (ideas, skills, facts) to be taught to a group of students. Examination syllabuses were usually exiguous and of little use in curriculum planning, most teachers preferring to refer to the questions contained in examination papers. Unfortunately, the term syllabus, although still widely used, tends to have the connotation of a static body of knowledge which seldom changes, as evidenced by GCE geography syllabuses, which in the twenty years following the inception of the GCE were hardly modified at all. Further, syllabuses seldom contained any aims or objectives though this is now being changed. For these reasons, it may be more appropriate to write of a geography curriculum document which would contain not only aims and objectives but suggested content and methods of evaluation. But the word "content" is itself ambiguous unless defined more precisely. By "content" I mean the principles, concepts, skills and attitudes which it is intended the students will learn. For example, Reilly's gravity model of the factors determining the level of interaction between two towns would be a principle, a fault-scarp would be a concept, drawing a proportional divided circle would be a skill and developing a concern for environmental quality would be an attitude. Most content cannot, however, be taught in a vacuum, it needs to be understood as referring to concrete situations. Thus the gravity model may be thought of in the context of the interaction between say Turin and Milan in Northern Italy, the fault-scarp in the context of the Great Rift Valley in Africa, drawing a proportional divided circle in the context of a comparative study of occupations of the populations of France, Germany and the United Kingdom, and concern for environmental quality in the context of a study of the urban development of Glasgow. I would, therefore, use the term "context" to describe the factual information derived from particular areas which illustrate the principle, concept, skill or attitude.

Having clarified the terminology to be used, let me now try to indicate how the curriculum planning process might operate. This process is best seen as operating at two levels, the general and the specific. At the general level, the content of courses are outlined for the whole school or institution. This is illustrated in Figure 7.1. This shows the process as operating as a system in which there is a sequence of operations over time. First, it is assumed that the school or college will have decided what its broad educational aims are. As shown, this may be affected by the output of thinking within the philosophy of education which itself will be affected by society's demands upon the educational system, though these may act much more directly on schools. Given that the overall aims of education have been determined, then decisions have to be made about the school curriculum which will help to achieve these aims; decisions for example as to whether geography should feature on the curriculum and, if so, what kind of geography. If it is decided to teach earth science within the natural sciences' department, then the geography to be taught may only be human geography. But even then it may be necessary to be more specific

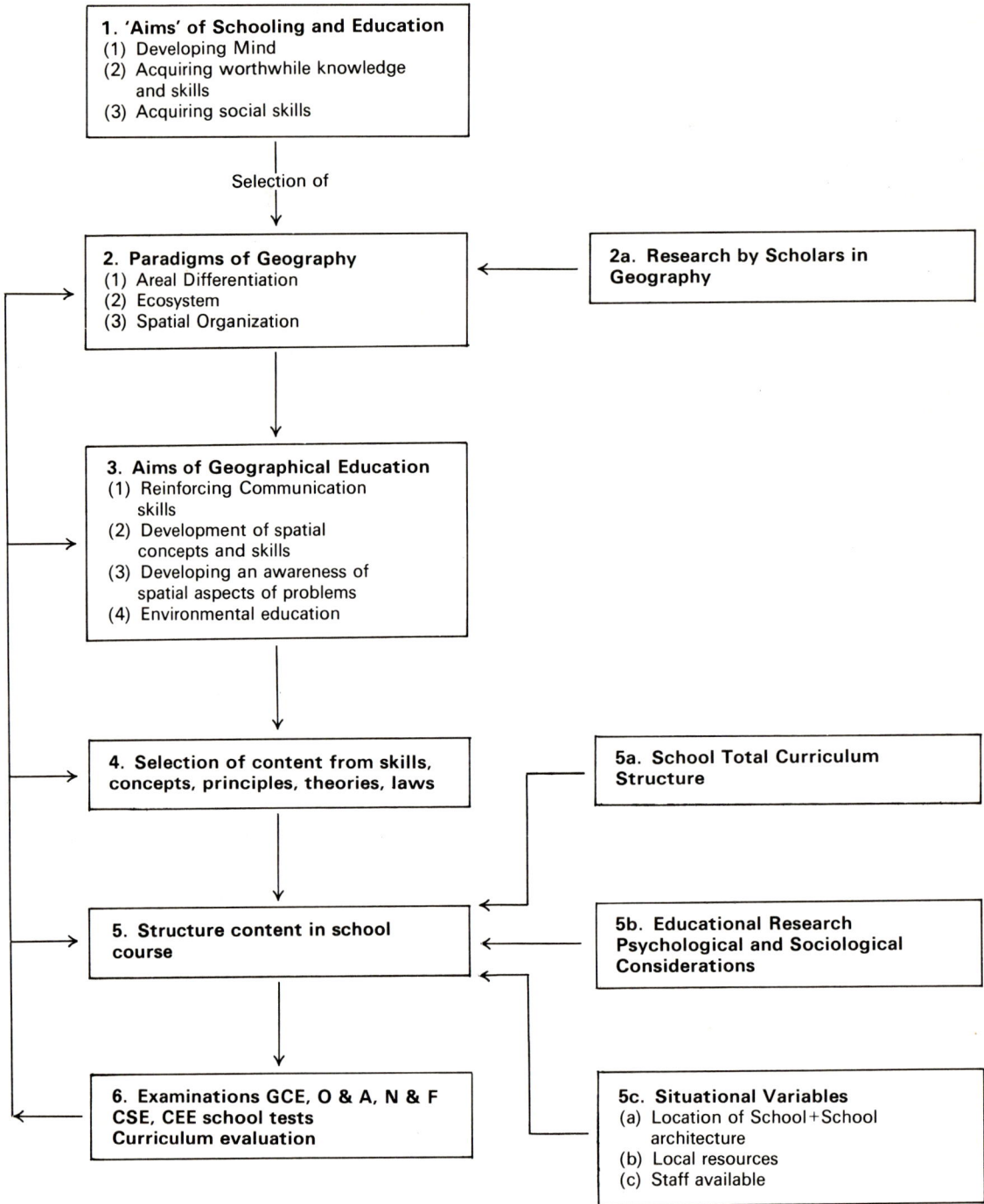

FIGURE 7.1. Model for curriculum planning in geography at the general level

as to what aspects of human geography should feature on the curriculum. The range of geography discussed in Chapter 2 makes this a pertinent question. Though I personally favour an ecosystem approach for schools, this view is not shared by all and the GYSL project, for example, chose a much more limited view of geography. As indicated in Figure 7.1, decisions about this need to be made not only in the light of what academic geographical research has to offer, but also in the light of curriculum theory.[71] Once the type of geography to be taught has been decided then the aims of geographical education may be formulated. Again, I have indicated what my own views on these aims are in Chapter 3 in relation to the ecosystem model of geography. These aims, in conjunction with the model of geography adopted, will help to determine the content to be selected. The sequencing of the content must be influenced partly by the logical structure of the subject; for example, teaching the concept of fault-scarp presumably must be preceded by the teaching of the concepts of scarp and fault. But sequencing will also be affected by the knowledge we have from psychological research in education about the level of difficulty of particular ideas. Broadly, we may divide concepts into two broad groups using the terms devised by the American psychologist Robert Gagné:[72] (1) concepts by observation, namely those it is possible to acquire through experience, though we need language to name them; examples would be sea, river, hill, slope, wood; (2) concepts by definition; that is, concepts which have been devised by man to serve his intellectual needs, but which could in no way be acquired by observation, for example, the location quotient illustrated in Chapter 4 is a simple idea, but it is man made and cannot be "discovered" through experience of the landscape or of economic life; it is a concept which has to be taught to be learned. The distinction is vital, because whereas most concepts by observation may be learned when pupils are still in the piagetian stage of concrete operations, concepts by definition are best taught when students are able to think in a hypothetico-deductive way. However, within this two-fold category there are subdivisions. Concepts by observation may be more or less difficult depending on how near or how far from the pupil's experience a particular concept is; a beach may be a simple concept to an English schoolboy,

a much more difficult one to a boy from Botswana. On the other hand, a concept like a continent is concrete, but is difficult to experience, because of its size, to any pupil. Similarly, in the category of concepts by definition, the concept of "density of population" is one which relates two variables, population and area, and as such is relatively simple; on the other hand the concept of "minimum cost location" for a factory is one which is dependent on a large number of variables, or in the field of meteorology the concept of an adiabatic lapse rate is one relating changes in air temperature with height in conditions in which the air concerned is neither losing nor gaining heat from its surroundings.

The content selected and sequenced according to the process indicated above will yield a list of principles, concepts, skills and attitudes which will be arranged in groups according to the years in which they are being taught and under subheadings according to the type of geography being taught. For example, using the ecosystem model of geography, I have used such headings as "geomorphological system, climatic system, biotic system, agricultural system, manu-facturing system and settlement system". Teachers will choose headings which are appropriate to their purposes. I would argue that it is not necessary in most cases to specify the context in which the content is to be taught, unless this is required by the examinations for which candidates are being entered. The advantage of not specifying the context is that it leaves teachers much more free to devise teaching units drawing upon a wide range of resources. It may legitimately be argued, on the other hand, that by not specifying the context, pupils and students may acquire a fragmentary view of the world in which they live and that it would be wise to ensure that examples were chosen from the developed capitalist world, the socialist world, and the developing nations.

Lastly, in the process of curriculum planning at the general level comes the evaluation of what has gone on. This is something which occurs both during the process of curriculum planning (formative evaluation) and after it has taken place (summative evaluation). As indicated in Chapter 5 it may occur through student and teacher appraisal, and the assessment of students both formally and informally. It should result in a review of all aspects of the curriculum at the general level. Clearly, the changes which may occur from year to year may be minor though on occasions a strong shift of views may result in a major change in aims and content. The system should be seen to be open to change. Although the head of the department will bear a major responsibility for this process, all members of a department should be involved since they will have to put the planned curriculum into effect.

At the specific level, the task is that of planning specific teaching units, based on the overall design for content given by the planning at the general level. This is a task that all teachers are involved in and they will be familiar with the process. It may be formalized as in Figure 7.2.

Given the content to be taught in any given year, there is still the need to indicate more precisely what kinds of objectives are appropriate to a particular group of students, and for the teachers who are to undertake the work. The level of the objectives will be determined by what teachers believe can be achieved with a particular group of pupils. If anything, it is best not to underestimate what pupils can achieve. Again, teaching strategies, as was seen in Chapter 4, are influenced by the objectives chosen, by the resources available and again by teacher and student characteristics.

Teachers tend to adopt the teaching strategies which they believe will fit into their personal style. Once the teaching unit has been put into action then evaluation inevitably occurs and modifications can take place next time it is used. Perhaps the objective was too simple, perhaps the teaching strategy adopted did not work (or worked particularly well), perhaps more up-to-date resources were required. It may be that what happens during a teaching unit indicates that some of the content is not worth including, or depends on content not yet taught. Hence there is feedback to the general planning level as well.

In a large department, teaching units will tend to be produced co-operatively in the sense that the kind of unit to be produced and the context of the ideas to be taught will be decided at departmental meetings. The putting together of the materials and the procedures suggested will be undertaken for any one unit or groups of units by an individual member of the department. This method of producing units enables the special knowledge and aptitudes of each member

to be used to the benefit of the whole department, since once a teaching unit is available it can be used by all members of the department. It may also encourage experimentation, since some members may find themselves using procedures which they would not have designed themselves, and in so doing, come to appreciate other teaching strategies.

It may be appropriate here to indicate briefly the role of the head of department and the functions of each member of staff in a secondary school geography department. In a small department arrangements with respect to teaching duties, to resources, to examinations, to curriculum renovation, may be informal and depend on the interaction between two or three teachers who have perhaps got to know one another very well. In a large department in a comprehensive school, the situation is a little more complex. In the first place the number of teachers teaching geography may be five or six or even more. In the second place, only a minority of those teaching geography may be specialists in the sense of having a degree in geography or a Certificate in Education in which geography was a main subject studied. In the third place, because of the large numbers of students being taught, there will be a need to monitor what is going on in various classes to prevent classes being too much out of step with one another. Consequently it may be necessary to formalize the organization of the department.

What does this imply? First, regular meetings of all those teaching geography at least once a term, but possibly once a month, during which departmental problems may be hammered out and departmental policy determined. Second, the head of department may do well to allocate specific responsibilities to various members of staff. In general it may be argued that the more experienced the members, the greater the responsibility that should be given them. For example, one member might be asked to be responsible for textbooks, another for non-book printed materials, yet another for audio-visual aids and so on. How a department manages its resources will depend on the school policy with respect to these. Some schools will have a large resource centre common to many subject areas with a media resources officer in charge. Others will have resources scattered throughout the school in subject stock rooms; in all probability a

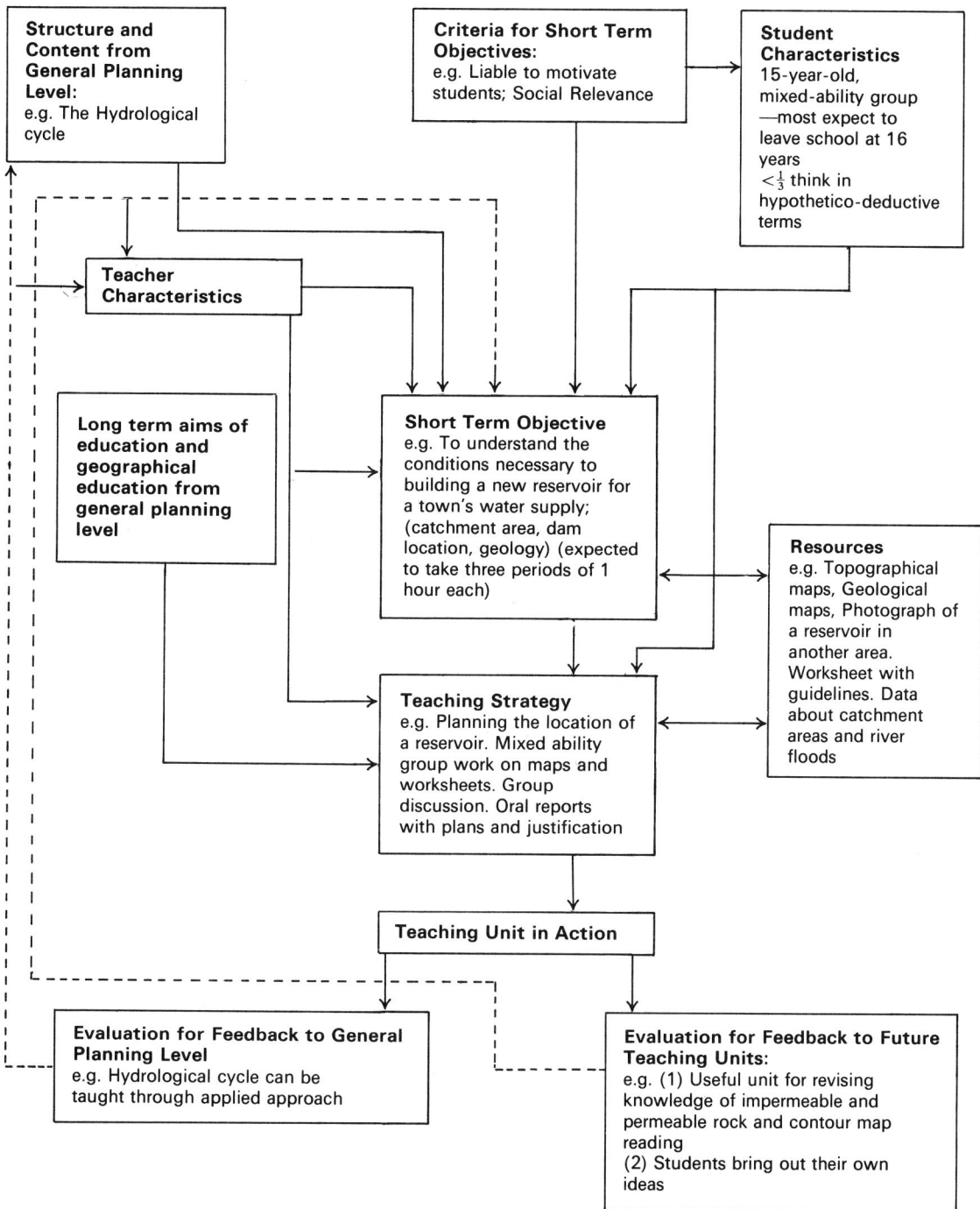

FIGURE 7.2. Model for curriculum planning in geography at the instructional level

number of resources specific to geography will be kept within the department, such as fieldwork instruments, large-scale and medium-scale maps, aerial photographs and various work sheets. The third implication is that the geography curriculum will need to be spelt out in some detail so that all staff members are clear as to what has to be done. Since some members will need more guidance than others, it will be useful for a large number of teaching units to be readily available which structure closely what teachers and pupils are expected to do. These can be accumulated over the years by a department, various members of staff having contributed individual units. It will be important for these units to be brought up to date from time to time, or even scrapped altogether if they are no longer relevant. Similarly new units will be added as these are developed by individual teachers. Inevitably the brunt of the work will be borne by those members of the staff who are specialist geographers.

The ordering of resources will normally fall on the head of department, but what is ordered will be discussed among staff members. The choice of text and other books may be a difficult task if money is tight. It may not be possible to spend much money on supplementary topic books, and it may be necessary to concentrate on basic texts. But the choice of such a text is a delicate matter since it may have to last for years. It must therefore be the kind of book which is more concerned with making principles explicit rather than concentrating on factual information which is likely to get out of date.

The head of department will need to give special attention to the most recently appointed members of his staff to help them in what is a difficult stage in their careers. This involves making sure they understand what is expected of them in their teaching but at the same time giving them support on matters of discipline, with guidance on lesson planning, with resources and generally keeping up their morale when it is flagging. It may well be that a new member of staff, like a student teacher, may be allocated to a "counsellor" from among the experienced geography staff.

7.2 CURRICULUM DEVELOPMENT

Curriculum planning becomes curriculum development as soon as the system operates to produce improvements in the geography curriculum from time to time. Many would argue that this is the only workable form of curriculum development since it arises out of the experience of teachers in and out of the classroom and is related to actual teaching situations. Nevertheless, various organizations and particularly the Schools Council have deemed it necessary to stimulate curriculum development by injecting into the education system ideas and materials for the improvement of curricula. I am conscious that in writing about curriculum improvement, questions are being begged since the nature of an improvement is not specified. One can only return to fundamental questions about the purpose of education (the development of mind, the development of abilities and skills useful to society), and ask whether the proposed changes in the curriculum satisfy such criteria. The unstated assumption is that all curriculum development has such aims in view.

Although curriculum development projects in geography go back in the United States to the early 1960s when the American High School Geography Project was launched under its director Nicholas Helburn (1964–70), in Britain the Schools Council launched its first geography projects in 1970: the Geography 14–18 project and the Geography for the Young School Leaver (GYSL) project. Both were aimed at secondary schools, but whereas Geography 14–18 was concerned with the impact of the so-called "new geography" on schools and particularly with average and above average ability children, GYSL was developed to cater for the students who would be expected to leave school at 16 years of age without taking examinations. In time, the GYSL project team came to see their courses as being suitable for a much wider range of ability than the so-called "less-able" school leaver. Thus at present GYSL courses are offered for CSE and O-level students as well as for the "young school leavers" who were the original target group. Each project adopted a somewhat different style of curriculum development. The Geography 14–18 team spent some time in analysing the situation they were faced with including the nature of new developments in geography. The team decided that no curriculum development would take place unless teachers were convinced of its need and unless the examination system were involved, since in Britain teachers are very much influenced by the demands of the examiners. Consequently the team attempted to work as change agents with a limited number of schools by interacting with members of geography departments as well as by suggesting certain teaching units and teaching procedures. Their pamphlet *A new professionalism for a changing geography*,[73] was a useful analysis of the developments in the geography teaching–learning system as they saw it. They also arranged for a new O-level Geography Examination to be available through the Cambridge Local Examination Syndicate. This process of curriculum development was necessarily slow since the team could not directly interact with more than a small number (10) of schools.

In due course the number of schools involved increased and at the time of writing (1979), about 100 schools are working to develop their geography curricula on lines first outlined by the Geography 14–18 project. A *Handbook for school-based curriculum development in geography*[74] was published in 1977.

The GYSL project was more concerned with producing a course and resources which would quickly become available to teachers, many of whom would not necessarily be specialist geographers. Consequently, although the team involved in the project worked closely with teachers in schools, the thrust of their efforts was directed to handing over to teachers teaching units which they could use, if necessary without further modification, rather than towards getting teachers to produce their own improved curricula and teaching units. Thus by 1974 Nelson had published the first theme, *Man, Land and Leisure*, and the others, *People, Place and Work* and *Cities and People* followed in fairly rapid succession. In terms of the number of teachers taking up the resources produced, this proved a most successful project, though the quality of the work undertaken with the project resources may vary considerably. Little by little, the GYSL and the Geography 14–18 projects began to come closer together in their operation. GCE O-level and CSE examinations were devised around the GYSL project themes, and

the materials available as a result of the Geography 14–18 project[75] began to take on the appearance of a structured course. Similarly, the GYSL co-ordinators encouraged teachers to produce their own resources.

The History Geography Social Science 8–13 project which began its operations in 1971 adopted a similar view to the Geography 14–18 project about the role of curriculum teams, namely that of encouraging school-based curriculum development. Nevertheless, the project also produced exemplar materials which are available to teachers. Those in comprehensive schools where the curriculum of the first two or three years is organized on an integrated basis, may find the project's ideas of use.

The latest project to be financed by the Schools Council is the Geography 16–19 project based on the University of London Institute of Education. The aim of the project is to help teachers develop their sixth-form geography curricula for all sixth formers, whether for those doing traditional A-level courses or for those on other courses. The team, benefiting from the experience of previous projects, realized that though the aim of turning teachers into curriculum developers is a good one, the process needed structuring fairly tightly so that teachers were given sound guidance in their task. Thus the project team had many discussions with teachers all over the country and produced a set of aims, a framework document in which was stated the kind of geography which they thought would be appropriate (a man–environmental model); they outlined major themes which would form the broad categories within which the content would be arranged; they arranged for the University of London School Examinations Council to sponsor an A-level examination based on the project's ideas and they are arranging for the publication of teaching units produced by the team and by the co-operating schools.[76] The project's structure is outlined in Figure 7.3.

Thus curriculum development in geography by agencies outside the school has evolved in the direction of those agencies acting in co-operation with schools so that what is proposed is seen to be realistic to teachers. At the same time, curriculum development teams have realized that if they are to help teachers to develop their geography curricula, they must provide more

Identify Needs of 16–19 Year-Olds

Identify the Contribution of Geography to the 16–19 Curriculum

Survey the Current Situation in 16–19 Geography

Outline the Broad Aims for 16–19 Geography

Define the Framework for a Core Geography Curriculum in Terms of Concepts, Skills and Values for all 16–19 Year-Olds

For the One Year 'New Sixth Former'

Outline Criteria for Selection of Objectives and Content for Units of Work

Devise Appropriate Teaching and Learning Approaches and Materials

For the 'A' Level Group

Teach and Evaluate in Schools and Colleges

Either As Sample Units Within Present Syllabuses

Or As Part of New Examination Courses Giving Scope for Development of Assessment Techniques

SURVEY

AIMS AND OBJECTIVES

DEVELOPMENT

EVALUATION

DISSEMINATION CONTINUES

AUTUMN 1976

AUTUMN 1977

AUTUMN 1978

AUTUMN 1979

SUMMER 1980

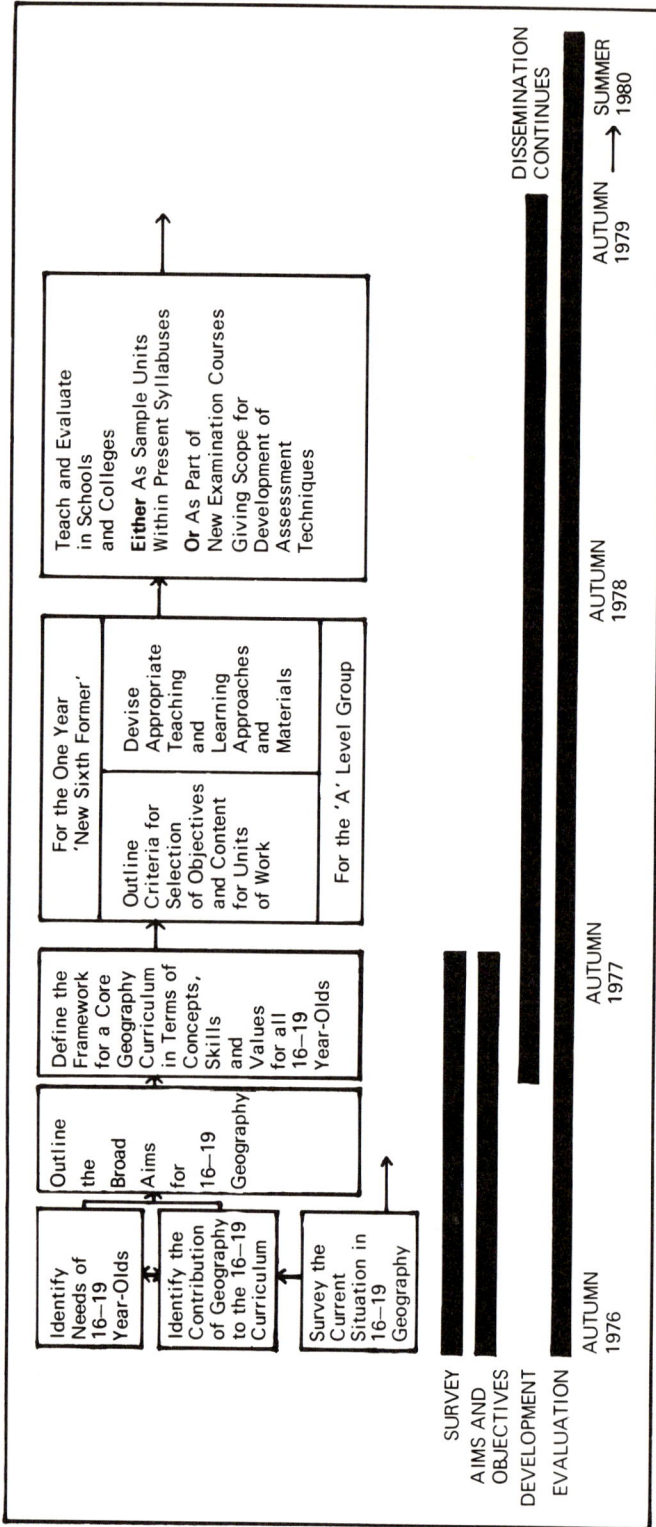

© *Copyright Schools Council Publications 1977.*

FIGURE 7.3. Structure of the Geography 16–19 Curriculum Development Project.

than stimulating ideas; teachers have a limited amount of time to spend on this task and firm guidance with suggested schemes of work and curricula frameworks will be seen as useful help rather than interference in the curriculum planning process.[77] Teachers are still free to adapt or even reject the schemes put forward by curriculum development teams. Research into the take-up of "new geography" by teachers seems to show that, on the whole, there is considerable willingness to experiment with new curricula, though inevitably the older ones among us are more reluctant than the younger ones.

Appendix: Source Books and Brief List of Organizations providing Educational Materials

SOURCE BOOKS

Abraham, H. J. *World problems in the classroom.* Paris: UNESCO, 1973.

Bates, J. N. *Further developments in geography teaching: a 1979 bibliography.* Sheffield: Geographical Association New Techniques and Methods Working Group, 1979.

Berry, P. S. *Sourcebook for environmental studies.* London: George Philip, 1975.

Brewer, J. G. *The literature of geography,* 2nd ed. London: Bingley, 1978.

Council for Environmental Education. *Directory of centres for outdoor studies in England and Wales,* 2nd ed. London: Council for Environmental Education, 1973.

Department of Education and Science. *The environment: sources of information for teachers.* London: HMSO, 1979.

Geographical Association. *Bibliographic notes.* Produced irregularly from 1979. (Topics include: Population geography at A-level; Assessment and the geography teacher; Mixed ability group work in secondary school geography.)

Graves, N. J. (ed.) *New Unesco source book for geography teaching* London: Unesco/Longman (in press).

Hancock, J. C. and Whiteley, P. F. (compilers) *The geographer's vademecum of sources and materials,* 2nd ed. London: George Philip, 1978.

Harris, C. D. *Annotated world list of selected current geographical serials in English, French and German.* University of Chicago Department of Geography, Research Paper No. 137, 1971.

—— *Bibliography of geography.* University of Chicago Department of Geography, Research Paper No. 179, 1976.

Laffin, J. *New geography, 1970–71.* London: Abelard-Schuman, 1971.

Lock, C. B. M. *Geography and cartography.* London: Bingley, 1976.

Long, M. *Handbook for geography teachers.* London: Methuen, 1974.

Geographical Dictionaries and Gazetteers, for example:

Paxton, J. *Statesman's year book world gazetteer.* London: Macmillan, 1979.

Stamp, L. D. and Clark, A. N. (eds) *A glossary of geographical terms,* 3rd ed. London: Longman, 1979.

Statistical sources and guides to, for example:

CBD Research Ltd. *Statistics Europe; Statistics Asia and Australia; Statistics Africa* etc. Beckenham: CBD Research Ltd.

Central Statistical Office. *Guide to official statistics.* London: HMSO, 1978.

Fullard, H. (ed.) *Geographical digest.* London: George Philip, published annually in May.

United Nations. *UN Statistical yearbook.*

ORGANIZATIONS

This if a brief list only. Other organizations providing educational materials may be found in, for example, *The geographer's vademecum of sources and materials.*

British Airports Authority, 2 Buckingham Palace Gate, London SW1.

British Airways, Victoria Terminal, Buckingham Palace Road, London SW1W 9SR.

British Broadcasting Corporation, School Broadcasting Council, The Langham, Portland Place, London W1A 1AA.

British Gas Corporation, 326 High Holborn, London WC1V 7PT.

British Rail
LMR Euston Road, London NW1 1HT.
SR Waterloo Station, London SE1 8SE.
WR Paddington Station, London W2 1HA.

British Steel Corporation, Information Officer, 151 Gower Street, London WC1E 6BB.

Centre for World Development Education, 128 Buckingham Palace Road, London SW1.

Commission of the European Communities, 20 Kensington Palace Gardens, London W8 4QQ.

Commonwealth Institute, Kensington High Street, London W8 6NQ.

The Conservation Trust, 246 London Road, Earley, Reading Berks, RG6 1AJ.

La Documentation Photographique, 13 Rue du Four, 75006 Paris, France.

Embassies of various countries will often provide educational resources; addresses in the London telephone directories.

The Geographical Association, 343 Fulwood Road, Sheffield S10 3BP.

George Philip Group, 12–14 Long Acre, London WC2E 9LP.

High Commissions of Commonwealth countries; addresses in the London telephone directories.

Independent Broadcasting Authority, 70 Brompton Road, London SW1.

Information and Documentation Centre for the Geography of the Netherlands, Geografisch Institut van de Rijksuniversiteit, Heidelberglaan 2, Utrecht, Netherlands.

Japan Information Centre, 9 Grosvenor Square, London W1.

National Audio-Visual Aids Centre, 254 Belsize Park, London NW6.

National Coal Board, Hobart House, Grosvenor Place, London SW1 7AE.

Ordnance Survey, Dept No. 32, Romsey Road, Maybush, Southampton SO9 4DH.

Petroleum companies often supply educational aids; addresses in London telephone directories and *The Geographer's vademecum.*

Port of London Authority, World Trade Centre, Europa House, East Smithfield, London E1.

Royal Town Planning Institute, 26 Portland Place, London W1N 4BE.

The Schools Council, 160 Great Portland Street, London W1N 6LL.

Town and Country Planning Association, 17 Carlton House Terrace, London SW1Y 5AS.

Unesco, 7 Place de Fontenoy, 75700 Paris, France.

References and Notes

1. Hollins, T. H. B. (ed.) *Aims in Education.* Manchester University Press, 1964.
   Hirst, P. H. and Peters, R. S. *The logic of education.* London: Routledge and Kegan Paul, 1970.
   Holly, D. *Beyond curriculum.* St Albans: Paladin, 1974.
2. Department of Education and Science. *Education in schools: a consultative document.* Cmnd 6869. London: HMSO, 1977.
3. See Young, M. F. D. (ed.) *Knowledge and control.* London: Collier-Macmillan, 1971.
4. James, C. *Young lives at stake.* London: Collins, 1968.
   Warwick, D. *Integrated studies in the secondary school.* University of London Press, 1973.
   Adams, A. *The humanities jungle.* London: Ward Lock Educational, 1976.
5. Hirst, P. H. 'Liberal education and the nature of knowledge', In Archambault, R. D. (ed.) *Philosophical analysis and education.* London: Routledge and Kegan Paul, 1965.
6. See Chapter 4 in Graves, N. J. *Geography in education.* London: Heinemann Educational, 1980.
7. King, A. R. and Brownell, J. A. *The curriculum and the disciplines of knowledge.* New York: Wiley, 1966.
8. Kuhn, T. S. *The structure of scientific revolutions,* 2nd ed. University of Chicago Press, 1970.
9. Relph, E. *Place and placelessness.* London: Pion, 1976.
10. Chorley, R. J. and Haggett, P. (eds) *Frontiers in geographical teaching,* 2nd ed. London: Methuen, 1970.
11. Chorley, R. J. and Haggett, P. (eds) *Models in geography.* London: Methuen, 1967.
12. Alber, R., Adams, J. S. and Gould, P. *Spatial organisation: a geographer's view of the world.* Englewood Cliffs, N. J.: Prentice-Hall, 1971.
13. Schon, D. A. *Beyond the stable state.* London: Temple Smith, 1971.
14. Smith, D. M. *Human geography: a welfare approach.* London: Arnold, 1977.
15. Peet, R. (ed.) *Radical geography.* London: Methuen, 1978.
16. Harvey, D. *Social justice and the city.* London: Arnold, 1973.
17. Saarinem, T. F. *Perception of the drought hazard on the Great Plains.* University of Chicago, Department of Geography Research Papers, No. 106, 1966.
18. Tuan, Yi-Fu. *Space and place.* London: Arnold, 1977.
19. Downs, R. M. and Stea, D. *Maps in minds.* New York: Harper and Row, 1977.
20. Balchin, W. 'Graphicacy', in Balchin, W. (ed.) *Geography: an outline for the intending student.* London: Routledge and Kegan Paul, 1970, pp. 28–42.
21. Martin, G. C. and Wheeler, K. (eds) *Insight into environmental education.* Edinburgh: Oliver and Boyd, 1975.
22. Tolley, H. and Reynolds, J. *Geography 14–18: a handbook for school-based curriculum development.* Basingstoke: Macmillan Education, 1977.
23. Sockett, H. *Designing the curriculum.* London: Open Books, 1976.
24. Marsden, W. E. 'Principles, concepts and exemplars and the structuring of curriculum units in geography', *Geographical Education,* 2 (1976) 421–9.
25. Bloom, B. S. (ed.) *Taxonomy of educational objectives, handbook 1: Cognitive domain.* London: Longman, 1956.
26. Krathwohl, D. R., Bloom, B. S. and Masia, B. B. *Taxonomy of educational objectives, handbook 2: Affective domain.* London: Longman, 1964.
27. Eisner, E. W. 'Instructional and expressive educational objectives: their formulation

and use in curriculum', in *Curriculum Evaluation*, AERA Monograph, No. 3. Chicago: Rand McNally, 1969.

28. Resources for this theme were obtained from the Japanese Information Centre, 9 Grosvenor Square, London W1.

29. Information derived from Graves, N. J. and White, J. T. *Geography of the British Isles*, 5th ed. London: Heinemann Educational, 1978.

30. Resources from 'La documentation photographique', No. 6028 (1977).

31. Guest, A. *Man and landscape*. London: Heinemann Educational, 1974.

    Ebdon, D. *Statistics in geography*. Oxford: Blackwell, 1978.

    McCullagh, P. *Data use and interpretation*. London: Oxford University Press, 1974.

32. Catling, S. 'Cognitive mapping exercises as a primary geographical experience', *Teaching Geography*, 3 (1978) 121–3.

33. Walford, R. *Games in geography*, 5th ed. London: Longman, 1975.

    Haigh, J. M. *Geography games*. Oxford: Blackwell, 1975.

    Dalton, R. *et al. Simulation games in geography*. London: Macmillan, 1972.

34. For a discussion of this area see Graves, N. J. *Geography in education,* op. cit., Chapter 9 and Graves, N. J. (ed.) *New Unesco source book for geography teaching*. London: Longman (in press), Chapter 10.

35. A discussion of the whole concept of evaluation in curriculum development is contained in Tawney, D. (ed.) *Curriculum evaluation today: trends and implications*. London: Macmillan, 1976.

36. For a brief reference to this see Graves, N. J. 'Changes in attitude towards the training of teachers of geography', *Geography*, 63 (1978) 75–84.

37. Bailey, P. *Teaching geography*. Newton Abbot: David and Charles, 1974.

38. Boden, P. *Developments in geography teaching*. London: Open Books, 1976.

39. Graves, N. J. *Geography in education,* op. cit.

40. Hall, D. *Geography and the geography teacher*. London: Allen and Unwin, 1976.

41. Marsden, W. E. *Evaluating the geography curriculum*. Edinburgh: Oliver and Boyd, 1976.

42. Manson, G. A. and Ridd, M. K. (eds) *New perspectives on geographic education: putting theory into practice*. Dubuque, Iowa: Kendall-Hunt, 1977.

43. *Teaching Geography* is published by Longman for the Geographical Association, Subscription £9.60 per year; *Geography* is published by the Geographical Association, 343 Fulwood Road, Sheffield, S10 3BP; subscription £9.30 per year. Subscription to both journals is £14 per year (1980/81 rates).

44. *The Journal of Geography* is published by the National Council for Geographic Education at the University of Houston, Houston, Texas 7704, USA. Cost $18 per year including membership.

45. *Geographical Education* is published by the Australian Geography Teachers' Association, Sydney Teachers' College, Camperdown, 2050 Australia.

46. *Classroom Geographer* is published by Brighton Polytechnic, Falmer, Brighton.

47. Barrow, R. *Common sense and the curriculum*. London: Allen and Unwin, 1976.

48. Hirst, P. H. and Peters, R. S. *The logic of education*. London: Routledge and Kegan Paul, 1970.

49. Peters, R. S. *Ethics and education*. London: Allen and Unwin, 1966.

50. White, J. P. *Towards a compulsory curriculum*. London: Routledge and Kegan Paul, 1973.

51. *The Journal of Curriculum Studies* is published by Taylor and Francis Ltd, 10–14 Macklin Street, London WC2B 5NF.

52. *The British Journal of Educational Studies* is published by Basil Blackwell and Mott Ltd, 5 Alfred Street, Oxford OX1 4HB.

53. Department of Education and Science. *Teaching of ideas in geography*. HMI Series, Matters for Discussion, No. 5. London: HMSO, 1978.

54. Graves, N. J. *Curriculum planning in geography*. London: Heinemann Educational, 1979.

55. Haggett, P. *Geography: a modern synthesis,* 3rd ed. New York: Harper and Row, 1979.

56. *Geography for the young school leaver*. Sunbury, Middx.: Nelson, 1974–8.

57. Tolley, H. and Reynolds, J. *Geography 14–18: a handbook for school-based curriculum development*. op. cit.

58. Blyth, A. *et al. Curriculum planning in history, geography and social science.* Bristol: Collins, 1976.

59. Marsden, W. E. *Evaluating the geography curriculum.* op. cit.

60. Tawney, D. (ed.) *Curriculum evaluation today: trends and implications,* op. cit.

61. Hamilton, D. *et al.* (eds) *Beyond the numbers game.* Basingstoke: Macmillan, 1977.

62. Deale, R. N. *Assessment and testing in the secondary school.* Schools Council Examinations Bulletin, No. 32. Andover: Evans/Methuen Educational, 1976.

63. Graves, N. J. *New movements in the study and teaching of geography.* London: Temple Smith, 1972.

64. Salmon, R. B. and Masterson, T. H. *The principles of objective testing in geography.* London: Heinemann Educational, 1974.

65. Kurfman, D. (ed.) *Evaluation in geographic education.* Belmont, California: Fearon, 1970.

66. Hones, G. H. 'Objective testing in geography', *Geography,* 58 (1973) 29–37.
Marsden, W. E. 'Analysing classroom tests in geography', *Geography,* 59 (1974) 55–64.
See also Okunrotifa, P. O. *Evaluation in geography.* Nigeria: Oxford University Press, 1975.

67. Clarke, E. (ed.) *Objective and completion tests in O-level geography.* London: John Murray, 1975.

68. See for example *Geography,* 65 (1980) 60–64.

69. A fuller treatment of this whole area may be found in Graves, N. J. *Curriculum planning in geography,* op. cit.

70. Department of Education and Science. *Aspects of secondary education in England.* London: HMSO, 1979.

71. See Lawton, D. *Social change, educational theory and curriculum planning.* University of London Press, 1973.

72. Gagné, R. M. 'The learning of principles', in Klausmeier, H. J. and Harris, R. M. (eds) *Analysis of concept learning.* New York: Academic Press, 1966.

73. Hickman, G., Reynolds, J. and Tolley, H. *A new professionalism for a changing geography.* Bristol: Schools Council Geography 14–18 Project, 1973.

74. Tolley, H, and Reynolds, J. *Geography 14–18: a handbook for school-based curriculum development,* op. cit.

75. See Schools Council. *Geography 14–18 Project:* Units on Industry, Urban geography, Transport networks, Population. Basingstoke: Macmillan Education, 1979.

76. For information on this project see *Geography 16–19 Project News,* available from the Geography 16–19 Project, University of London Institute of Education, Bedford Way, London WC1 0Al.

77. Further information on the contribution of three Schools Council projects to secondary school geography can be obtained from Rawling, E. (ed.) *Geography into the 1980s.* Sheffield: Geographical Association, 1980 (in press).

Index

Figures in italic refer to illustrations